MODEL RAILWAYS

MODEL RAILWAYS

THE COMPLETE GUIDE TO DESIGNING, BUILDING AND OPERATING A MODEL RAILWAY

CYRIL FREEZER

NEW BURLINGTON BOOKS

A QUINTET BOOK

Published by New Burlington Books
6 Blundell Street
London, N7 9BH

Exclusive to Coles in Canada

ISBN 1-85348-333-8

This book was designed and produced by
Quintet Publishing Limited
6 Blundell Street
London N7 9BH

Creative Director: Terry Jeavons
Designer: Annie Moss
Project Editor: Damian Thompson
Editor: Lydia Darbyshire
Illustrator: Danny McBride
Photographer: Ian Howes

Typeset in Great Britain by
Central Southern Typesetters, Eastbourne
Manufactured in Hong Kong by
Regent Publishing Services Ltd.
Printed in Singapore by
Tien Wah Press (Pte.) Ltd.

CONTENTS

INTRODUCTION

With a model railroad it is possible not only to simulate the operation of a full-sized railroad but also to create a small, ideal world, which you can visit whenever you are in need of relaxation. Moreover, it can be a challenging hobby, providing problems that tax your skills and ingenuity to the full – problems that, although every bit as real as those you may face in your everyday life, are not important. You can leave them aside indefinitely, or you can even ignore them altogether and try something different. With a model railroad you can have all the excitement of the struggle and all the satisfaction of success, but none of the anxieties.

Railroad modelling has been around for just over a century, for it was in the 1880s that the toymakers of Nuremberg began to produce a complete range of equipment – locomotives, coaches, wagons, tracks, stations and a host of accessories – from which a realistic replica of a 'real' railroad could be assembled. A little later, largely through the work of the British company Bassett-Lowke and the efforts of Henry Greenly, the disciplines of model engineering and the concept

BELOW

Bill Roberts' model of Prien am Chiemsee, Germany, is a superb evocation of modern Deutsche Bundesbahn practice and demonstrates what can be done to create a realistic representation of an actual station in N gauge.

BELOW

In the beginning was 1 gauge, 10mm scale on 45mm gauge track, which in pre-World War 1 days was the most popular size. Today it is a very specialized scale, and yet the hobby has grown so much in the intervening years that it now has a larger following than in its heyday. Märklin have reintroduced models in 10mm scale and this picture shows how impressive the models are at eye level.

of scale modelling were added to the tinplate train base. Because 'scale modelling' sounds better than 'playing trains', this aspect of the hobby has always been emphasized, even to the point at which some enthusiasts decry the commercially made, ready-to-run model. This is about as sensible as climbing on to a roof and then kicking the ladder away, for the ready-to-run model is not only the point at which most people enter the hobby, but it can be a considerable help in many aspects of 'serious' railroad modelling.

The great thing about model railroads is that, apart from the reduced availability of overseas locomotives, the same principles apply the world over. Measurement is a minor problem, for while most of the world uses the metric system, the US remains wedded to feet and inches. The UK has, for the past 60 years, used a peculiar mixture, largely because it was realized at an early stage that when building small models, a millimetre is much more convenient than an inch as the basic unit of measurement. In this book, dimensions are given in both systems, mainly on the basis of 25mm to the inch. Occasionally, however, a metre is treated as a yard for we are not dealing with precise measurements, but rather suggesting the appropriate size to aim for. The only area where precision is needed is in the track gauge, which is dealt with in the appendix. Here, metric standards have been used exclusively since, even where these are expressed in inches, the figure is almost always a direct conversion of a metric standard.

CHAPTER ONE

THE CHOICE BEFORE YOU

Once upon a time it was simple and straightforward. A model railroad was as faithful a copy of the local full-sized system as could be constructed in the space available, and it was worked according to the public timetable, suitably cut down to fit the limitations of the system. But that was a long time ago.

Today, the choice facing a railroad modeller is vast. A casual visit to a large modelshop or a brief glance through the advertising pages of any of the main model railroad magazines will show that a gigantic selection of models, covering many countries and ranging from historic steam to ultra-modern diesel and electric prototypes, is available. There are also many scales and gauges on offer, ranging from the extremely large gauge 1 (45mm) down to the minuscule Z gauge (6.5mm). The choice is wide. Of course, if you decide to model the sugar-cane railroads of Fiji you will have a few problems, but none that cannot be overcome with a little determination and a fair amount of time, since the essential information is available, and some of the vital bits and pieces can be provided by a few specialist suppliers.

At least, that is one way of looking at it – from the viewpoint of the established enthusiast who not only has a lot of experience in the hobby but also knows a good deal about full-sized railroads. After all, before you can

RIGHT
Although we consider diesel power to be up to the minute and the latest development in the 165 year story of railroads, modern motive power has been well established for several decades. In this case the main locomotive, an Eastern Region British Rail Deltic is a model of an extinct type, whilst the diesel railcar's prototype is scheduled for early replacement.

FACING PAGE
Refuelling a Main Central GP38 on an HO gauge Burlington Northern layout. Note the amount of detail squeezed into a fairly small space on this most realistic model.

252 252

ABOVE

This N gauge Inter City set is a fine example of modern motive power in miniature, seen here operating over an N gauge layout of Geislingen, Germany.

consider modelling the sugar-cane railroads of Fiji you need to know that they existed in the first place. Likewise, before you can make an intelligent choice of scale, gauge and standards you need to know the advantages and disadvantages of each in relation to your own likes, dislikes and abilities. Which should come first? It appears an insoluble dilemma.

It is not, of course, an impossible problem. The simple fact is that in railroad modelling, as in most other activities, it is easier to go along with the crowd. Different approaches are possible, but many rely on the support of small specialist manufacturers, often one-man cottage industries. Alternatively, they require the modeller to build a good deal of the railroad in his or her own workshop. While these are practical propositions, they do involve a good deal more effort, and if, as so often, you look to your hobby as a means of relaxing after a hard day's work, this is the last thing you want.

Relying on the products of the major manufacturers does, to a large extent, limit your choice to two basic sizes – HO/OO, with 16.5mm gauge, and N, with 9mm gauge – for these are the ones that command the main support. The selection of models in HO is somewhat more extensive than it is for N, but even in N you still have the choice of most national systems in Europe and North America and the option of using the latest diesel and electric motive power or of looking back to the golden years of steam. At least, this is the immediate impression one gets from a study of the catalogues.

SCALE AND GAUGE

The twin factors of scale and gauge are fundamental to any model railroad. The key element is the gauge, the distance between the top inner edges of the rails, a dimension that is normally quoted in millimetres. Dimensional standards in railroad modelling have been metric for over 50 years, although in the US dimensions are frequently converted to their imperial equivalents, which tend to result in some very odd sizes.

The scale is derived from the gauge – or at least it should be. However, as the full-sized track gauge is an odd size to begin with – 4ft 8½in (1435mm) – and the old German toymakers were still wedded to inches in the late 19th century, dividing 1¼in by 4ft 8½in not only proved difficult, but ended up at 17/64in. The adoption of metric standards simplified calculations and usually resulted in a round number of millimetres on the model representing 12in on the prototype. Since a round metric measure of 12in cannot easily translate into a round metric equivalent of a metre, Continental manufacturers and enthusiasts adopted the ratio 1:87, which is the same as 3.5mm to 12in, the scale of HO or, in the US, HO scale.

This ratio is steadily gaining ground because it is independent of the system of measurement used. You could even work in cubits, if you happened to be able to measure in cubits! More to the point, it is easy to use a pocket calculator to calculate the size of the model if the dimensions of the prototype are known. It is not even necessary to resort to mathematics, since scale rules, which allow the dimensions of the prototype to be laid directly on to the model, are available from specialist modelshops.

However, scale is of direct interest only to the scratchbuilder. If you are going to use ready-to-run models or work from kits, it is the manufacturers' responsibility to get the sizes of the parts correct. The better firms quote the scale used in their literature and on their packaging, which is a useful guide, and they also state the gauge employed, and it is this latter practice that has led to the term 'HO scale' gaining credence – it is really a shortening of phrase 'the scale for HO gauge'.

The newcomer, faced with plethora of scales and gauges, can easily get confused. The solution is to discover the name of the size of model you favour. Forget the rest, since they do not apply to you. Those who wish to delve further should refer to the appendix, where a more detailed description is given.

ABOVE

The Gateway Transportation Co.'s offices on an N gauge layout. An array of modern diesel motive power is displayed on the rear tracks, the most visible unit – the only one clearly identifiable – being a Santa Fe GP38-2 hood unit.

However, a visit to your nearest supplier may reveal a slightly different state of affairs. You can be reasonably certain that there will be good stocks of the current products of the major national manufacturers, together with models produced by a foreign company for your home market. You will not, however, necessarily find that everything listed is available at any given moment.

If you live a little off the beaten track you may not have a good modelshop close to hand. Even in densely populated Europe it is not unusual for modellers to find that there is no large stockist within a 40km (25 mile) radius of their homes and that they must either make a special excursion or rely on the products of their local high-class toy dealer. Here, the stocks will be smaller and possibly concentrated on the products of one or two major firms. If you are thinking in terms of prototypes (the full-sized originals) from overseas, you will probably have to face the fact that your main supplies will come from a few specialist concerns.

Although it would be wrong to suggest that, if you cannot get hold of the basic requirements for your hobby in your own town, you are in dire trouble, it certainly helps, in the opening phases at least, if the essential equipment is readily to hand. Certainly, the availability of suitable models is a very important consideration, and problems in this area are not easily dismissed. Mail order is, of course, a partial solution.

FACING PAGE

British Rail's diesel-electric 'Blue Pullman' sets, the forerunners of the modern diesel-powered HST, have disappeared from the prototype and although models were made for OO gauge, these too have now been discontinued. Many, however, remain in active service on layouts.

As a result, the favoured choice for the newcomer to railroad modelling remains much as it was in the beginning – that is, a model of a nearby railroad in more or less its current condition, with a nostalgic steam-age system, again based on the national system, as a second choice. The probability is that if you live in Britain you will favour OO, while if you live in the US, HO will be your choice. In both countries, N would be a second choice, particularly if space were limited. In Continental Europe the preference between HO and N seems to be less marked, and modellers are as likely to choose N as HO.

You may well, however, be saved having to agonize. it is indeed very probable that most of what you need to make a start is lying, gathering dust, on top of a wardrobe or up in the loft. A train set, an attractively boxed set of locomotive, rolling-stock, track

FACING PAGE

This HO gauge layout of Mudrock Valley Railroad shows a rake of reefers – refrigerated boxcars – lined up alongside a servicing depot. The high level walkway is provided to allow workers to load the cars with dry ice through hatches in the boxcar roofs. The inclusion of an essential servicing feature of this nature marks the differences between a well-thought-out model railroad and a highly developed train set.

and, in the better outfits, turnouts and a mains-powered controller as well, is not only the traditional entry into the hobby, it also contains the essential core of a beginner's system. Train sets are often ignored because they used to be extremely crude. Today, however, even the cheaper 'starter' sets that are aimed at children, contain models that are more accurate representations of the prototype than all but the best 'scale' models of 50 years ago.

Of course, the mix in the set may not be exactly what you want, but this is not important. One feature of a model railroad that is rarely appreciated is that the parts that do most to set the locale and the period are either the locomotives, coaches and wagons, which merely rest on the track, or the buildings, which, although more firmly connected to the landscape, are nevertheless individual items that can be taken away and replaced. The main infrastructure – the baseboards, track and electric wiring – are inherently neutral. The landscape – if indeed any is included – is representative of a few basic types of terrain that, within reasonable limits, could be anywhere. As a result, therefore, although you may start with a model based on your local prototype railroad, you can always modify this later on. Clearly, the more equipment you have, the more it will cost you to change, and so, at the outset, it is best to start in a modest way. Indeed, the real choice before you is to begin with a set that will allow you to change your mind if you wish at a later date.

ABOVE

A row of kit-constructed buildings is set in front of the tracks at Geislingen on this N gauge Deutsche Bundesbahn layout. Note that while the shops and houses firmly anchor the period and region of the scene, they are set below the track level, thereby affording an uninterrupted view of the trains and track.

ABOVE LEFT

Narrow gauge models are ideally suited for small spaces. Here we see a Festiniog (Wales) Railway 0-4-0 heading a short rake of quarrymen's coaches. Here the modeller has relied heavily on his own resources, building the models in the home workshop from kits, components and basic raw materials.

RIGHT

This model, running through part of the 3ft 6in gauge South African Railways, proved difficult to find. Most enthusiasts choose a prototype supported by the mass manufacturers, but for those prepared to take that extra bit of trouble, the world is open.

CHAPTER TWO

WHERE WILL IT GO?

A model railroad is probably the largest modelling project any amateur is likely to undertake. This is implicit in the fact that the prototype is measured in miles and, in general, in hundreds of these rather large units. Fitting this into a normal home can cause difficulties, and before we can even think about the layout itself, we must consider the possible sites within the home so that we can begin the design process knowing that we can turn our plans into reality.

The ideal solution is a separate railroad room, but, like most ideals, that is not all that easy to put into practice. A spare room may seem a good site, but it is a fact that the demands on family accommodation can increase over the years. It is probably better to look at non-habitable rooms – garages, cellars, attics and sheds. Fortunately, there is a lot of room under a permanently positioned railroad baseboard, and so a room can function as both a home for a layout and a store.

A basement is ideal for our purpose. It must be reasonably dry, for water is the main enemy of a model railroad. Similarly, outdoor structures need to be rigorously checked beforehand, not so much for rising damp, which rarely reaches baseboard level, but for the integrity of the roof. If there is any doubt, have it checked beforehand. One delightful layout came to a sad end when roofing contractors applied hot tar to a flat roof and sufficient fell through to wreck the model.

An attic is a good place for a model railroad. Stretching over the entire house, it can usually provide a reasonably large space, and, because access is generally up through a hatch near the centre, there are fewer problems in getting to the layout. An attic is not, however, without its problems.

One of the main hazards is variations in temperature. In winter the thermometer can go below zero; in summer the place can resemble an oven. It is not unduly difficult to instal additional insulation inside the roof when you line the rafters. This has the added advantages of improving the insulation of the home, thus reducing the fuel bills, and at the same time making it far less likely that you will have any trouble with frozen water pipes. The provision of a small, thermostatically controlled heater will eliminate any risk of frozen pipes and provide a reasonable amount of comfort when you are working on your layout in winter. It should also be remembered that tungsten lamps are a good

FACING PAGE

Despite the apparent size of this section of layout, the floor space needed is minimal, since the sheer rock face along which the track clings is mainly in the vertical plane and makes excellent use of the gap between floor and ceiling! This scene was taken on the Rensselader Model Railroad Society's HO gauge Berkshire & Western club layout.

BELOW

Ken Ashberry's Ashdown & Midport is a fine example of a compact 4mm scale layout, situated in a very small spare bedroom.

BELOW

Tregarrick, UK, on the imaginary North Cornwall Minerals Railway was built on a series of small baseboard modules and erected in a small cottage.

ABOVE
A popular solution to the space problem is the terminus-fiddle yard scheme, where, for example, the trains run off into a set of hidden sidings representing the rest of the system. Exit is often through a tunnel, and this photo shows a Deutsche Bundesbahn railcar set emerging into 'reality' from one. Despite the small area involved – roughly 400mm (16in) square – there is a feeling of space about the model.

source of heat while they are throwing light on the model scene, and much of the warmth needed during wintertime can be supplied in this fashion.

For most of the year, however, the problem will be to keep the roof space cool. It is possible, of course, to fit a skylight, but you may be unwilling to cut into a sound roof. This is, in any case, a job for a good firm of specialist contractors who will have had ample experience in fitting watertight skylights, a job that is outside the scope of the majority of builders. A simple solution is to provide a fan to suck the hot air from the apex of the roof and blow it out through the eaves. If you decide to do this, you must provide trunking, although this need only involve boarding over the space between two rafters. It is not important to seal the outside of this space, for the object is to get rid of the air, not to direct it into a specific location.

Installing blowers to direct cool air at the operator(s) is a more challenging task. It requires one specific skill, the ability to glean the necessary equipment from scrapyards and surplus dealers. Old fan heaters can be used as effective blowers, and a good deal of corrugated aluminium and plastic trunking ends up in rubbish skips every day. To the ingeniously minded DIY enthusiast, this is a challenge to relish.

The garage is another very obvious location for a permanent layout. It is big enough to meet most requirements yet not so large as to make the model burdensome. Garages are, in theory, provided as homes for family cars, but they are all too often so full of junk as to leave no room for the vehicles, which more often than not stand outside on the drive. On the other hand, it is perfectly possible to arrange the contents of a normal garage in such a way that it not only holds a good many miscellaneous and necessary items and *two* model railroads, but also leaves ample room not only for the car, but the lawn mower and grass trimmers as well. It has to be a small car, of course.

Over the years railroad modellers have come up with ways and means of getting a worthwhile model railroad into the living space in the home. To explain this, let us begin with a common starting point, the model railroad in a youngster's bedroom. All too often, the initial idea is to put a single, solid baseboard into one corner. However, the snag is that it requires a good deal of room. A baseboard measuring 1.8 × 1.2m (6 × 4ft) is rather large – as big as a double bed – yet, in model railroad terms, it is a trifle cramped. A relatively narrow railroad that runs around the walls is a better idea. Such a set-up has little effect on the overall space in the room, since, in most cases, it runs over the furniture, and also provides a much larger length of main line.

It is often suggested that a suitable layout can be mounted on a board that will slide under the bed when it is not in use. This proposal clearly cannot have been tried in practice. Anything placed under a bed rapidly becomes covered with fluff, another major enemy of model railroads. Most modelshop proprietors are very adept at removing fluff from the interiors of model locomotives, for it is one of the major reasons model locomotives are brought in for repair.

Another interesting suggestion is that the baseboard be hinged against a wall. This can be an effective solution, although many who have tried it have abandoned the arrangement. There are two inherent problems. The first is that before the layout can be folded away, every loose item has to be removed. If this is not done, the items will fall off and end up in a fairly inaccessible place. This is enough of a nuisance, but a more serious consideration is that a hinged baseboard has to be stoutly built to withstand the twisting and moving involved and is, accordingly, fairly heavy. This weight has to be poised on its edge when not in use and is potentially unstable. Should it accidentally come down when someone is in the way, it could well cause injury.

Fortunately, there are easier ways of dealing with the business of fitting a layout into a home. If it is not desirable to run the layout permanently around the walls, a portable layout can be built. The best approach is to mount it on a number of small baseboards. These sections – between six and eight will be sufficient – are fixed together to provide a larger layout, and when they are not in use they can be stored away out of sight. We will deal with the constructional details in Chapter 5; for the moment the question of size and storage is paramount.

First and foremost, the overall size of a portable layout, and, therefore, of the sections from which it is assembled, is set by the place in which it is to spend most of its time. This may be a handy recess or a spare cupboard; in either case, the overall size will be limited. It is also probable that it will have to be carried through the house, which means it must negotiate a normal-sized door opening. In practice, a board 1.2 × 0.6m (4 × 2ft) is about as large a unit as most people can move about single-handed without too much strain. This is also a convenient size to take along a twisting corridor or up a flight of stairs with a bend in the middle without any risk that it will come into contact with the walls. This is not advisable, as it usually results in damage to the model and to the decor, so you are in trouble both ways. Even a board this large can be awkward. A board measuring 1 × 0.5m (3ft 3in × 1ft 6in) is easier to handle and has the additional advantage that it will fit more easily into the rear of the family car.

Single-handed transit is an important factor since it is difficult, if not downright impossible, to guarantee that a willing helper will be on hand at any time. Of course, at public exhibitions it is not unusual to see layouts built on modules measuring as much as 2 × 1m (6ft 6in × 3ft 3in), but in such cases a club or other organized group is responsible, and not only are there usually anything up to a dozen individuals on hand (four willing volunteers and eight pressed men), but the passageways and doorways are wider than those in the average home.

The small baseboard has a further virtue: it may be easily placed on a workbench or even an old coffee-table so that the builder is not only able to carry out the major initial modelling but can also add finer detail at leisure, often in the company of the family. In addition, it is easy to stand such a baseboard on end to attend to the sub-baseboard wiring and turnout mechanisms under optimum conditions. Furthermore, because the work is being carried out on a relatively small area, it is not a daunting task to bring the complete section up to a high standard of realism. When one section has been completed, it is, naturally, not merely desirable to maintain the level of craftsmanship but to improve on it on the next section. This is why many small portable layouts are extremely effective.

ABOVE

Where space is limited, an obvious answer is to adopt N gauge, the smallest size widely catered for by the trade. This shot of a viaduct on a German layout gives an idea of what can be achieved in a small space with this scale.

CHAPTER THREE

DESIGNING THE LAYOUT

ABOVE

Copying an actual station, as was done in this N gauge model of Geislingen, Deutsche Bundesbahn, does not necessarily simplify the design process, since it is usually necessary to compress some parts of the prototype to fit it into the available space. Apart from that, finding a suitable prototype that will fit is not exactly easy!

Although a few newcomers to railroad modelling try to design their own layouts, most copy a published plan because they appreciate that it is necessary to know a good deal about full-sized practice to know how best to arrange the tracks. Comprehensive layout plan books are published and provide a good deal of inspiration.

One error many novice modellers make is to believe that it is desirable to create a completely original track layout. This is practicable only if the layout is *extremely* large, for all the workable schemes that fit small- or medium-sized spaces have been built time and time again. Therefore, in this chapter six schemes are described that have been tested in this fashion and can be adopted in the knowledge that they have already given pleasure to many railroad modellers.

All these schemes are for HO/OO, 16.5mm gauge, and most incorporate fairly tight curves that will be suitable for current ready-to-run equipment. The plans are covered with a squared grid, representing 300mm for metric or 12in for imperial standards. Although there is a small difference between these nominal sizes, it is a matter of only 4.8mm and, even in the largest space shown, amounts to a difference of only 40mm (1.6in) in total. This is rather less than the tolerance built into the designs, for quite apart from the difficulty of enlarging the dimensions of a small-scale plan to full size, room sizes and track components also tend to differ in size.

All the schemes are relatively small and the majority are also very simple. Another error often made by beginners is to believe that only one layout will be built. This is extremely rare. Many layouts have a life of between 5 and 10 years, and even if a layout has had a continuous history stretching well beyond this, there will have been several major reconstructions *en route*. Although this usually implies either a change of site or an extensive reconstruction, there was one instance of a layout being completely rebuilt to virtually the same plan simply because, after some 20 years, it was realized that not only were parts of the model showing signs of age, but in the interim considerable improvements had been made to track and track components.

More to the point is the fact that it has been shown that in the long run it saves time and effort if a small layout is built in order to

FACING PAGE

This is model railroading on the grand scale, a section of the Pasadena Model Railroad Club's HO gauge Sierra Pacific lines. The bridge is a magnificent model in its own right: the steel spans and trestles are correctly braced and joined by gusset plates, leaving no doubt in the viewer's mind that this could carry the load produced by four Western Pacific GP7 hood diesels in multiple – and more!

ABOVE

Road traffic – such as it is – is held up whilst a Baltimore & Ohio 0-8-0 trundles a freight across the grade crossing on an HO gauge B&O Railroad layout. A wealth of fine scenic modelling, incorporated into a relatively small area, creates the illusion of reality.

LEFT

When extensive scenic development is wanted on the model, space must be provided for at the outset. You cannot include an effective section of landscape, such as this Alpine extravaganza as an afterthought.

establish the most suitable modelling techniques so that the trial and error period is confined to a disposable system.

All the layouts shown on these plans could benefit from additional space to allow for longer lengths of straight track and, above all, longer loops and sidings, not to mention easier curves. It is not, however, advisable to attempt to reduce them in size, since a cluttered layout is ugly and difficult to run.

For N gauge there are two possibilities. The first is to treat each square as measuring 150mm or 6in; however, you must remember that, where an operating well is shown, it cannot be materially reduced since one's waistline does not vary with a change of modelling scale. The alternative option is to build the layout to the size shown for the larger scale but to adjust the track spacing and add a couple of additional sidings. This will materially improve the appearance and capacity of the schemes and is really the best way to exploit the spatial advantages of the smaller scale.

PLAN 1 – BASIC OVAL WITH SIDINGS

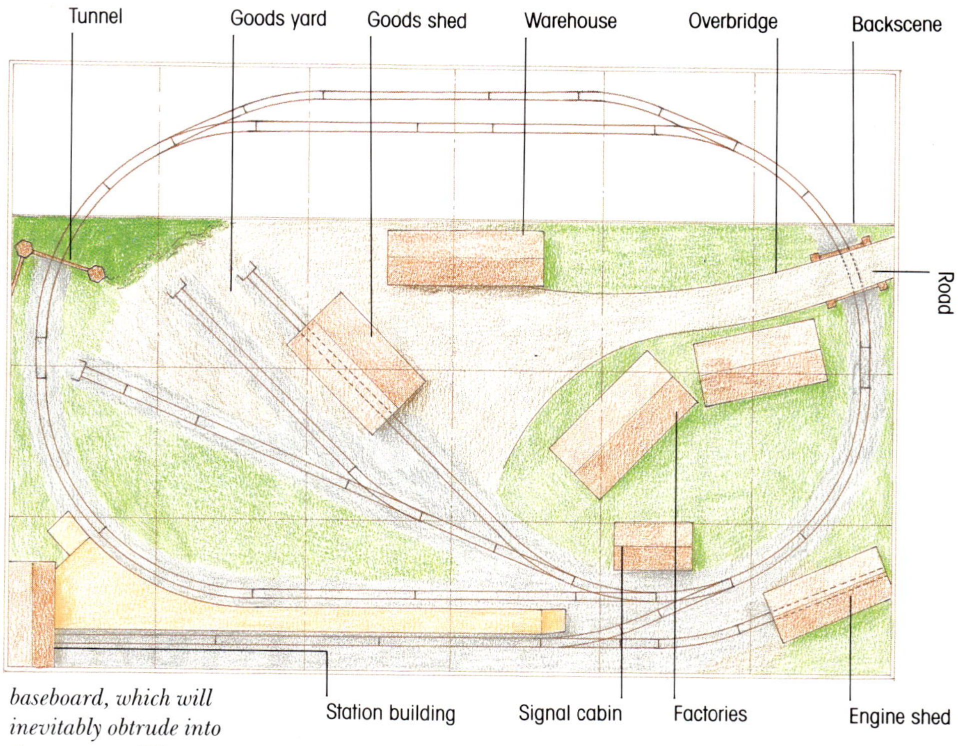

This simple scheme, which is on a 2.4 × 1.2m (8 × 4ft) baseboard, exploits an existing collection of sectional track and is capable of handling two trains – a goods train, which must circulate in a clockwise direction in order to shunt the yard, and a passenger train, which can be run either way. It can be worked intensively with only two locomotives, but a third will add to the interest.

A loop is situated behind the backscene. If a pair of curved turnouts is used, this can be as long as possible. It is hidden from view by the backscene, which is set roughly 300mm (12 in) from the rear.

The scenery suggests an urban setting with the tracks passing through road overbridges to disguise the inevitable hole in the backscene. As an alternative, the model could be set in a mountainous area and tunnel mouth substituted. Although it would be possible to hide the loops under a modelled mountain, this could create access problems when the layout was placed against a wall.

This scheme is a simple transition from the early train set oval, but the large baseboard, which will inevitably obtrude into the room, could be a problem.

PLAN 2 – SIMPLE SHUNTING LAYOUT

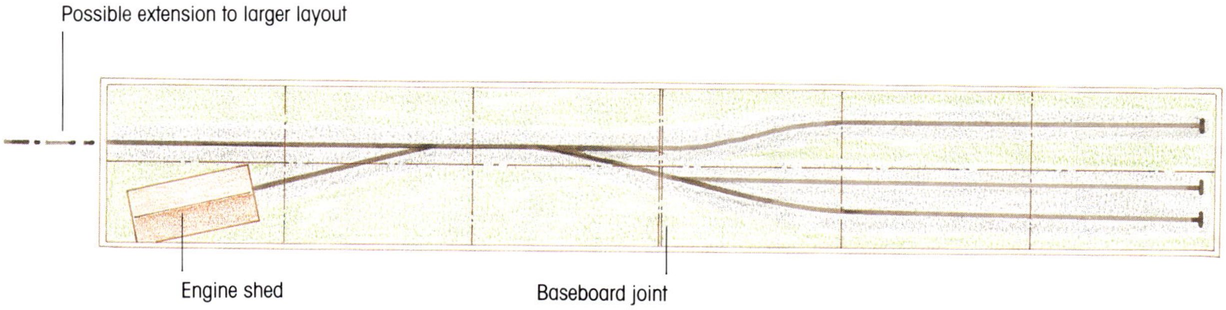

Where space is a problem, a small shunting yard, built on two relatively small baseboards, can provide a great deal of pleasure. This is a tried and tested scheme, which first saw light of day in the mid-1920s and has been copied on many occasions. It is suggested that the model is built on two 0.75 × 0.3m (2ft 6in × 1ft) baseboards, which are hinged along the central joint, the hinged sections being hidden by an overbridge. A second bridge disguises the point at which the track leaves the baseboard, and this will enable the model to be incorporated into a more ambitious scheme at a later date.

Although the basic construction of the layout can be completed fairly rapidly, there is considerable scope for detailed work within the confines of the relatively small area modelled. The layout is a first step away from the simple train set oval and all that it implies, and this makes it an ideal testing ground for modelling techniques. In N gauge, the entire scheme can be mounted on a single baseboard.

The operating potential is good, although limited. It requires only one small shunting locomotive and between 8 and 12 four-wheeled wagons. It is at its best when each wagon is distinctive in design or finish and thus readily distinguishable from its companions. In addition, operating the layout can be made more challenging if a series of cards, each labelled to correspond with a wagon, is shuffled and four dealt from the pack; the object is to assemble these four wagons into a train in the order of dealing. It is not as easy as it sounds.

PLAN 3 – SCENIC LOOPED EIGHT

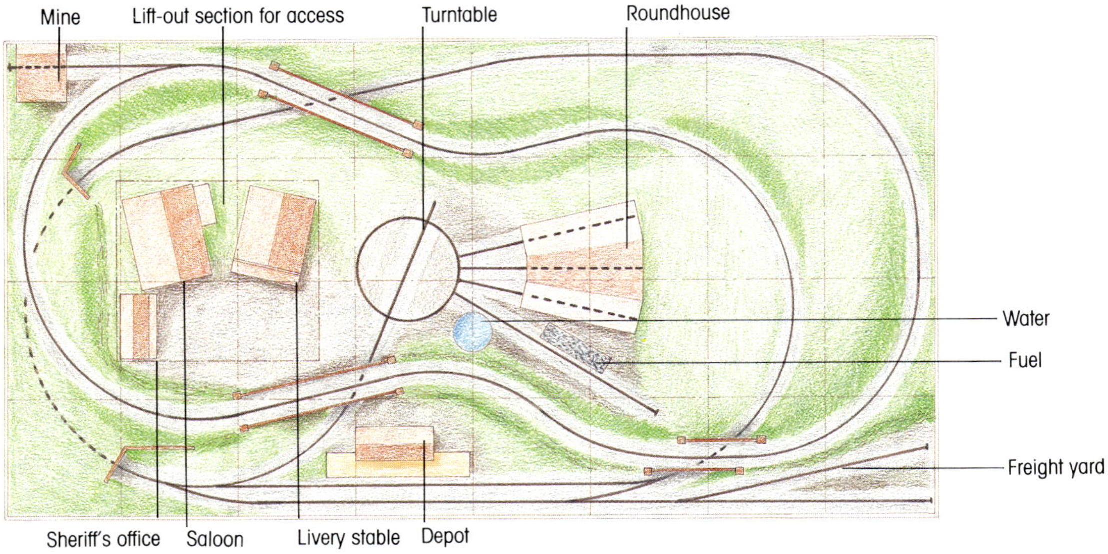

The 2.4 × 1.2m (8 × 4ft) baseboard fits snugly across the rear of most garages, and the exact length can be adjusted to meet the actual site. This makes it compatible with a medium or small car, particularly if the baseboard can be lifted high enough to clear the bonnet. It is also well suited to basement sites, where it is usually possible to gain access on all four sides.

The scheme illustrated is the popular 'looped eight', which provides a reasonable length of run in a relatively small area and, at the same time, creates an opportunity for some interesting scenic work. The setting is the State of Chaos, USA, the setting for all those B-movie westerns, for the sharp curves and relatively limited length of train are best suited to the old 4-4-0 and 2-6-0 locomotives and short, clerestory-roofed coaches. A roundhouse is provided to accommodate a small locomotive stud, and although this is highly unlikely in reality, it is extremely useful on the model, where it is probable that there will be rather more locomotives than are strictly necessary.

The scenic development suggests a mountainous region, with mineral deposits accounting for the mine buildings. The township served by the depot is assumed to be off the baseboard, but there is room for a few typical façades. Following the B-movie theme, there could be a saloon, a general store and the sheriff's office and jail.

If, as suggested, the layout is located in a garage, access to the rear tracks could be a problem; hence the provision of a removable access hatch.

PLAN 4 – MAIN-LINE OVAL WITH STORAGE ROADS

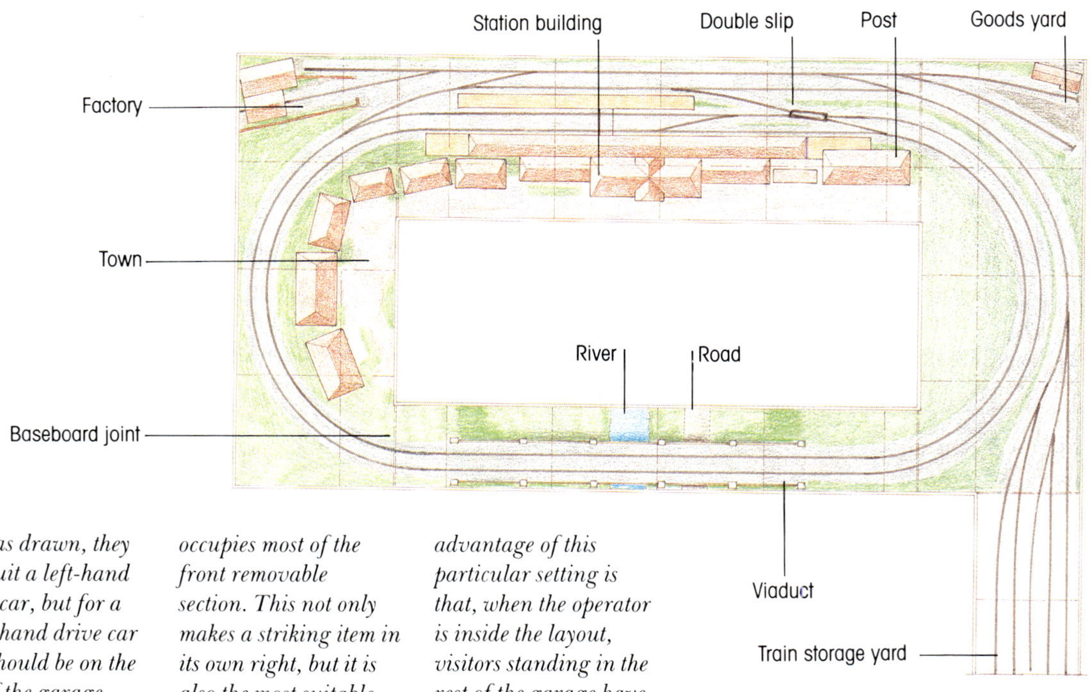

If, instead of a large solid baseboard, a 2.4 × 1.2m (8 × 4ft) layout is built on four sectional baseboards, not only is the resulting layout easy to dismantle and move, but there is a central well for the operator. Once again an end-of-garage location is envisaged, but the front section is removable so that a larger car can be housed in the garage.

The scheme follows European practice for intermediate stations and offers double-track operation. As storage capacity is limited, it is suggested that a series of storage sidings can be extended down one side; as drawn, they will suit a left-hand drive car, but for a right-hand drive car they should be on the left of the garage.

A long, straight viaduct is the principal scenic feature, and it occupies most of the front removable section. This not only makes a striking item in its own right, but it is also the most suitable spot on which to pose trains for inspection by visitors. One advantage of this particular setting is that, when the operator is inside the layout, visitors standing in the rest of the garage have an excellent unobstructed view of the trains.

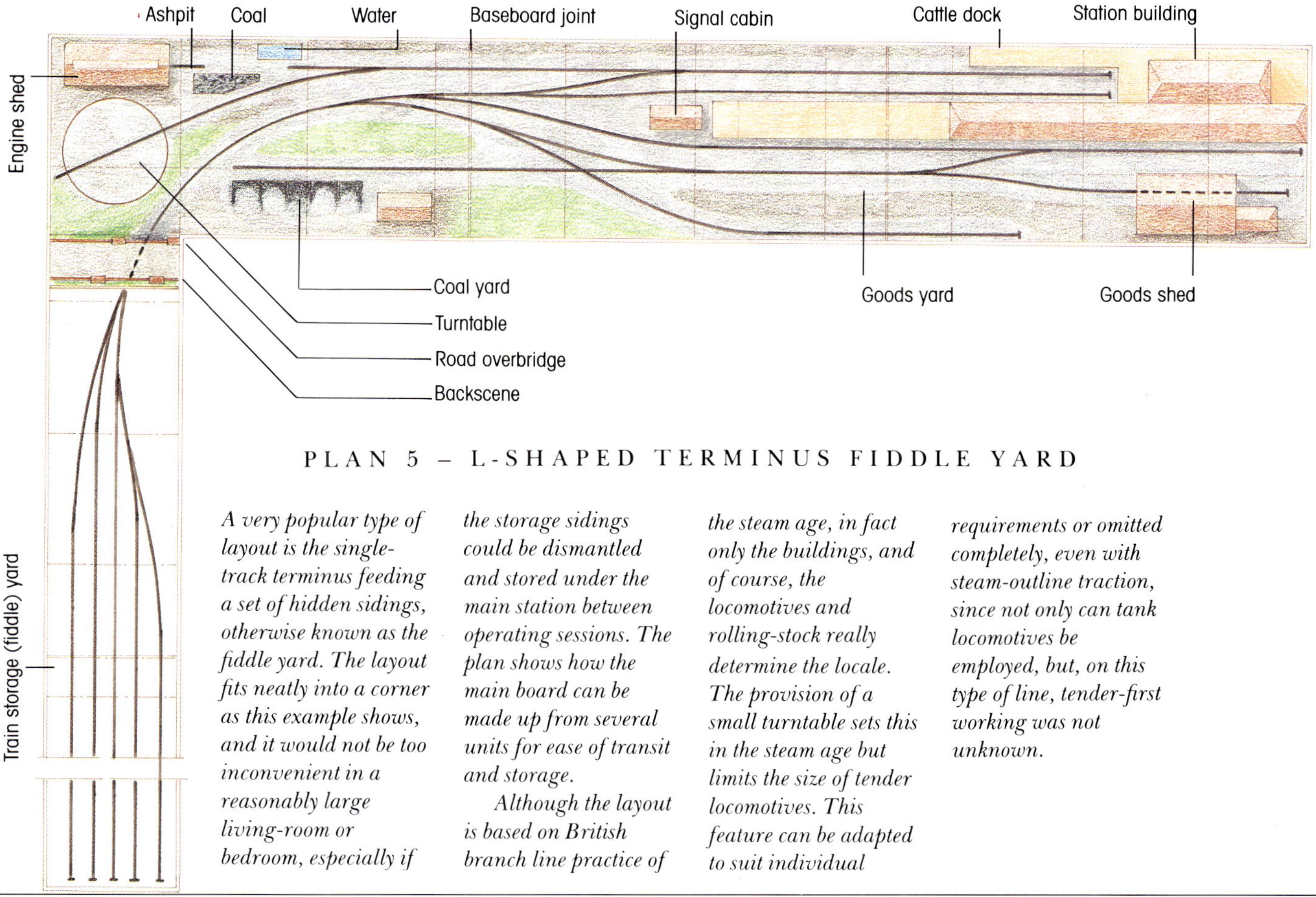

PLAN 5 – L-SHAPED TERMINUS FIDDLE YARD

A very popular type of layout is the single-track terminus feeding a set of hidden sidings, otherwise known as the fiddle yard. The layout fits neatly into a corner as this example shows, and it would not be too inconvenient in a reasonably large living-room or bedroom, especially if the storage sidings could be dismantled and stored under the main station between operating sessions. The plan shows how the main board can be made up from several units for ease of transit and storage.

Although the layout is based on British branch line practice of the steam age, in fact only the buildings, and of course, the locomotives and rolling-stock really determine the locale. The provision of a small turntable sets this in the steam age but limits the size of tender locomotives. This feature can be adapted to suit individual requirements or omitted completely, even with steam-outline traction, since not only can tank locomotives be employed, but, on this type of line, tender-first working was not unknown.

PLAN 6 – THROUGH TERMINUS IN SPARE BEDROOM

The final layout is for a small spare room measuring 2.5 × 1.9m (8ft 4in × 6ft 4in). It provides multiple-track running together with a fairly impressive main station and a small goods yard. The design is based on British steam-age practice, but, as the track layout has been considerably simplified to fit it into the space available, once again it is the structures and stock that set the scene.

This layout is about as complex a project as any novice could comfortably handle, yet it offers a very considerable operating potential. The provision of a small locomotive depot allows for the storage of rather more locomotives than are strictly necessary for the operation of the layout – most people like to have a surplus of motive power. Passenger trains start in either of the two long terminal roads and terminate in the outer through-platform, goods trains proceed from the sidings, run round the system and then terminate in the loop road. Although train lengths are limited, they are long enough to be interesting. This is a layout of which an N gauge version, in the same space, would really show dividends.

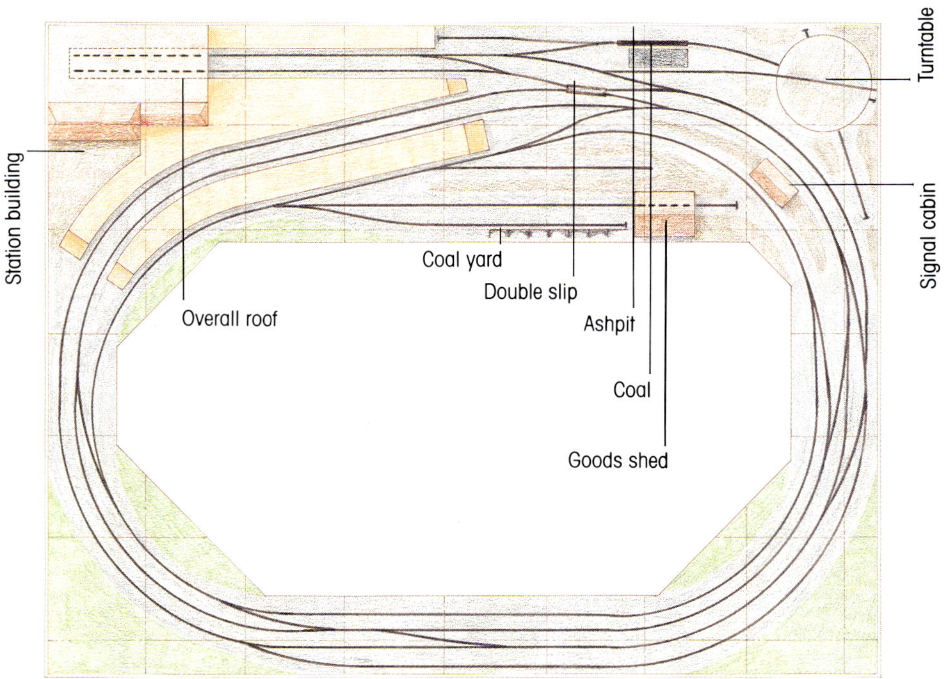

CHAPTER FOUR

LOCOMOTIVES AND ROLLING-STOCK

ABOVE

The German 2-10-0 *Kriegslokomotiv* was built in vast numbers to keep the traffic running during World War II. In post-war years, they turned up everywhere and were still running at the end of steam on the Deutsche Bundesbahn.

The locomotives and, to a slightly lesser extent, the coaches and wagons that go to make up the trains they haul are easily the most interesting part of most model railroads. Indeed, the layout itself can be regarded as no more than an elaborate show-case on which a selection of trains can be displayed. Whether the object is to run to a strict timetable or merely to sit and watch trains go by, the whole thing is meaningless without the star performers.

Logically, we should begin by planning a balanced collection of models to ensure that the end picture is a satisfying one. The first step should be to calculate how many locomotives are needed to work the service. Then the number of coaches and wagons required to provide the trains should be established. Then a specific prototype should be selected and the period in which the model is set and the exact locale of the layout determined. When all this was decided, you would then draw up a list of models you required and set about acquiring them. This idea had certain attractions in the past when it required much effort, in time and money, to build up a moderate-sized collection of models. It has less relevance today for not only do we have a wider choice but models are much more readily added to the collection, and so we can adopt a simpler strategy.

At the outset you need a test train. If, as is very likely, the first layout is being built around an existing train set, you have this already. If not, something cheap and cheerful is indicated, since this particular train will be roughly handled during construction. It is going to be derailed, it will probably fall on to the floor from time to time, it will be splashed with plaster and paint, and it will be

FACING PAGE

The roundhouse on an N gauge Cumberland Valley system shows a Norfolk & Western streamlined J class locomotive on the turntable. In front of it stands a Norfolk & Western Y6B 2-8-8-2 Mallet; left is a Pennsylvania Railroad K-4 4-6-2 Pacific, whilst a Fairbanks Morse switcher stands on the lead track.

BELOW

The classic American 4-4-0, used throughout the US for passenger and freight work – a simple, rugged design. This model is a coal burner and was owned by the Virginia & Truckee railroad.

ABOVE

The latest British electric locomotive, the class 91, has an unusual single-ended design. It is used solely on passenger services, with a DVT (Driving Van Trailer) at the other end of the train for reverse working. This 00 gauge model is in the new Inter City livery.

LEFT

LNER A4 Mallard – a ready-to-run steam locomotive by Hornby – moves on to the turntable, while an experimental diesel running in the late 60s on British Railways passes behind at the head of a train of maroon-liveried coaches.

picked up with dirty hands. It is the model equivalent of the contractor's train: useful while the line is being built but whisked out of sight before the grand opening ceremony.

Although you will not actually need the correct trains until the layout is reasonably complete and capable of serious operation, it is inevitable that you will begin collecting them before the first section of baseboard has got beyond the planning stage. By the time the layout is finished, the probability is that you will have somehow managed to acquire rather more stock than you can comfortably accommodate on the layout.

What we have in practice are two separate entities – a layout, which is a single unit, possibly forming a permanent structure built into a room, and a large number of individual models of locomotives, coaches and wagons that can run on the layout. Although the layout will limit the number of locomotives, coaches and wagons that can be run on it at one time, there is no theoretical restriction on the size of your collection, although limits are set by three factors of varying importance: the depth of your pocket; the amount of space you have to store the models; and the demands of realism.

BELOW

British Railways produced a range of standard steam locomotives in the 1950s, and this 00 model is a Class 4MT 4-6-0, used for secondary services. It is a good choice for a beginner's layout.

BELOW

The roundhouse shed, with numerous tracks radiating from a central turntable was, in the steam age, the main type of structure, except in Britain. This type of shed is still in use in Continental Europe to house both diesel and electric locomotives. This fine model formed part of the Fleischmann display at the Nuremberg Toy Fair for many years, demonstrating their remote controlled multi-road turntable.

Although there is nothing to prevent you amassing models simply because you like the look of them, your collection will be more satisfying if you follow a pattern. For a start, it helps to think in terms of trains. You may start with the train itself, indeed there are today some very good boxed sets, comprising a locomotive and matching coaches, which make this largely a matter of impulse buying. More probably, you will begin with a favourite locomotive and then collect the coaches and wagons it will haul. From there, you add extra locomotives that will also haul your coaches and wagons until you arrive, quite painlessly, at a nicely balanced collection that can work a sensible range of services over your miniature system.

In fact, the majority of model railroads do not need very many locomotives. Three well-chosen types can provide a realistic service, while anything over a dozen will probably become a slight embarrassment, because as soon as the storage capacity of the layout

STOCK STORAGE

The obvious way to store locomotives, coaches and wagons is to keep them in the makers' boxes. This is not only simple, but it also adds to the resale value in future years. If, for any reason the collection, or some appreciable part of it, is going to be untouched for any length of time, it is a good idea to store the individual boxes in larger card cartons and to seal the boxes and label them with a note of the contents. Alternatively, if you have any redundant chests of drawers, the individual boxes can be stored in these, but the existence of inexpensive flat-pack furniture should not be overlooked either – particularly when end-of-range items are being offered at knock-down prices.

Some enthusiasts like to display choice items from their reserve collection in glass-fronted cases of the kind produced by specialist firms and advertised in the leading magazines. Purpose-made storage boxes can also be bought, or made in the home workshop. These consist of shallow boxes with regular partitions that are lined with foam plastic. When lids and handles are added, these boxes form a convenient way of carrying the stock for a portable layout. If models have to be stored loosely in a box, they may be protected from damage by being wrapped in either acid-free tissue paper or sheets of padded kitchen paper. They can then be neatly arranged in layers, although additional protection should be provided by sandwiching a sheet of thick card between each layer.

ABOVE

A workhorse of the modern US railroad, a hood diesel, seen here in Union Pacific livery. This HO model is a good choice for a beginner, and can be pressed into any service.

is approached, operation begins to resemble a sliding-block puzzle. There is only one empty space, and, if you are to get a specific locomotive into it, you must first move everything else. There is, of course, a simple answer: do not have everything on the layout at once. Important though the trains and other rolling-stock are, they are not actually fixed to anything and can be taken off and replaced whenever you like.

It is customary to select locomotives, coaches and wagons that were almost certainly to be seen together, for this produces a more convincing overall picture. However, there is nothing to stop you introducing other trains into the scene, or even of setting out deliberately to use the layout as a show-case for trains from all over the world and from all periods of time. The important point is to

ABOVE

This scene is typical of the tracks in front of a British steam shed at any period up to the early sixties. However, the mechanical coal stage sets it at post-1930, whilst the locomotive insignia, the British Railways second Lion and Wheel totem, make it late fifties or early sixties.

RIGHT

This little HO gauge Deutsche Bundesbahn 2-6-01 was used on local and branch services. An ideal choice for a small steam age layout.

BELOW

The zenith of US steam power was represented by the Mallet articulated locomotives – two separate chassis under one enormous boiler, and built to haul mile-long freight trains across the continent. No one modelling a steam-age layout of any size can resist owning at least one of these behemoths.

LEFT

A loco depot is located in a corner of this HO gauge Hoosac Valley Railroad layout. An 0-8-0 switcher stands in front of a large timber-built coal stage close to the water crane which stands between two tracks. A two-road shed is in the background, while massed trees link the model to the backscene. Deceptively simple, the realism of this scene began with careful planning of the facilities.

BELOW

The Deutsche Bundesbahn V200 diesel hydraulic is one of the most successful designs of all time – rugged, reliable and widely used on non-electrified lines. Indeed no modern layout based in Germany should be without one.

keep each train consistent, so that each locomotive could have hauled the specific coaches to which it is coupled. The fact that it would be impossible to stand by the lineside and watch the *Flying Scotsman*, followed by the *Orient Express* and the *20th Century Limited* run past does not mean we cannot enjoy that spectacle on a model railroad. If one wishes to see a German *Kriegslokomotiv* pulling up at a station on the Southern Pacific line, that is the operator's privilege.

LEFT

Freight wagons are rarely as pleasing to the eye as the elegant locomotives that haul them (in this case a former London & North Eastern Railway pacific) – but they add a gritty realism to any layout.

CHAPTER FIVE

BUILDING THE BASEBOARD

The one thing that distinguishes a model railroad from a train set is the baseboard. The baseboard not only enables the modeller to lay the track in a realistic fashion, it also permits the wiring to be installed permanently and allows for the gradual addition of the hundred and one small details that help to create a convincing scene. It is the foundation of the model, and it is essential that it should be constructed with care, as it should be capable of lasting for a considerable time without sagging or distorting in any way. This is not quite as onerous a task as it sounds, for it is only what is expected of a piece of furniture.

Initially, baseboards bore a strong resemblance to workbenches. They were made from thick timber nailed to stout bearers, which were supported at regular intervals by massive legs. This is possible only where the layout can be permanently erected in a suitable room, where top-quality timber is both cheap and easily obtainable, and, ideally, where it is easy to find a reliable jobbing carpenter to build it for you. Even when all these conditions can be met, however, the system has a serious flaw in that it is inflexible. Not only would it be difficult to modify the arrangements at a later stage, but, should it be necessary to move, the entire railroad would have to be broken up and all the work that had been put into it would be wasted. What is needed is a system whereby the layout can be dismantled and re-erected without too much difficulty.

A popular method is to build the model on a series of frames that can be fixed together to form the whole. This not only produces a layout that can be dismantled without too much difficulty, it also allows one section to be removed for repair or reconstruction. Such an arrangement is, of course, essential if the layout has to be of a portable nature and enables it to be taken to a public exhibition. It has even been claimed that this system of construction allows a layout to be easily dismantled when moving house and quickly re-erected in the new home. This assumes, however, that a near-identical space will be available in the new house, and this is rarely the case.

TOOLS AND METHODS

Very few tools are needed to construct a baseboard – a tenon saw, square, screwdriver, hammer and a drill are essential. A workbench is useful, but an old wooden kitchen chair and a couple of G-clamps can suffice. The ideal fitting is a folding workbench, one of the few 'amateur' tools that have been accepted by professionals without reservation. A mitre-block is a useful acquisition to ensure accurate joints, and although it might be regarded as cheating, professionals use them.

The most useful power tool is, of course, the drill. The next in importance is the power jigsaw (sabre saw), a very versatile tool that is invaluable for cutting out trackbases. The hand-held circular saw is of less value and is potentially dangerous, although if it is housed in a sawbench it means that materials can be cut into accurate strips. Remember that, when you make the first cuts in a large sheet, you will need a helper.

Screws require pre-drilled holes, and while the countersunk and pilot hole are optional, the former makes for a neater finish while the latter speeds insertion and eases the workload. Special drill bits, which cut all three shapes in one pass, are available and save a good deal of time.

For gluing, PVA woodworking adhesives are preferable. Hot-glue guns are not at their best when used on large areas and are more appropriate for fitting smaller sections in place – for example, the cleats on the risers on open-top or L-girder framing.

A selection of G-clamps can be very useful, while an inexpensive corner clamp, of the kind sold for picture framing, is a handy device for holding two timbers at right angles while you screw them in place.

RIGHT

A section of the San Diego Model Railroad Museum's HO gauge layout, showing how neatly a baseboard edge can be finished by the provision of a deep fascia board, neatly curved to provide a 'soft edge' to the model. Note on the extreme right how local controls have been neatly let into the fascia. The train is headed by a selection of modern diesel locomotives, two SD40T-2s, an SD45 and a U28C.

BRIDGE 335.56 CALIENTE CREEK

The basic unit consists of a rectangular frame supporting a top surface on which the tracks are laid, and the construction is extremely straightforward, as the diagrams show. The framing is cut from 45 × 20mm (2 × ¾in) timber, preferably using a mitre-block to ensure that the cuts are square, and the joints are secured with screws and the top surface fixed in place.

The top is made from a piece of man-made board. When this type of construction was first adopted, the only readily obtainable material was plywood, which is still a very popular surface although it tends to be a trifle noisy. Chipboard has its adherents; it is cheap and rather less prone to drumming than ply. The harder types of insulation board, which will take fine pins easily, are also popular, but they require intermediate support. Hardboard is unsuitable, however, for not only is it difficult to pin or screw tracks and other fittings to it, it has an annoying tendency to twist and buckle.

It was rapidly realized, however, that although a solid top to the framing is ideal around the station, in the open countryside the fact that 'bedrock' lies immediately below the tracks prevents a realistic section of landscape from being created. The solution is the open-top baseboard. Here, the tracks are laid on a narrow trackbase, which is supported a convenient distance above the framing, this is explained in the diagram, and the method of using the dropped baseboard to permit the modelling of valleys is also shown. It will be seen that the tracks are carried on a relatively narrow trackbase, supported by risers secured to cross-members. Small cleats of 20mm (¾in) square timber are provided at the top of the risers so that the trackbases can be screwed to the risers. The open space is subsequently filled with landscape, buildings, roads and rivers, the modelling of which will be discussed in later chapters.

JOINING BASEBOARDS

There are two schools of thought about joining baseboards. The high-tech school use a variety of ingenious gadgets, the best of which undoubtedly are pattern-maker's dowels, which are designed to fit two halves of a pattern together with high precision. The low-tech school considers that, because absolutely precise alignment is extremely difficult to achieve and next to impossible to maintain, when wood is used as a base, a small amount of play is desirable so that final adjustment can be carried out by eye. This allows inexpensive, easily obtainable fittings to be used. One of the most popular is the coach bolt, preferably combined with a wing nut. Two of these on each joint allow baseboards to be linked in a semi-permanent fashion.

Assembly and re-assembly is a fairly slow and tedious process, so for portable layouts split hinges are a popular alternative. The hinge pin is taken out and replaced by a close-fitting pin – a wire nail is ideal, once the point has been filed to a smooth curve. The doctored hinge is then screwed across the joint while the baseboards are held in alignment with G-clamps. To separate the units, the pin is withdrawn, and both sides can be dealt with in less than 10 seconds. Make a few spare pins in case of loss., If top-quality hinges are used, this type of fitting is remarkably accurate.

OPEN-TOP BASEBOARD CONSTRUCTION

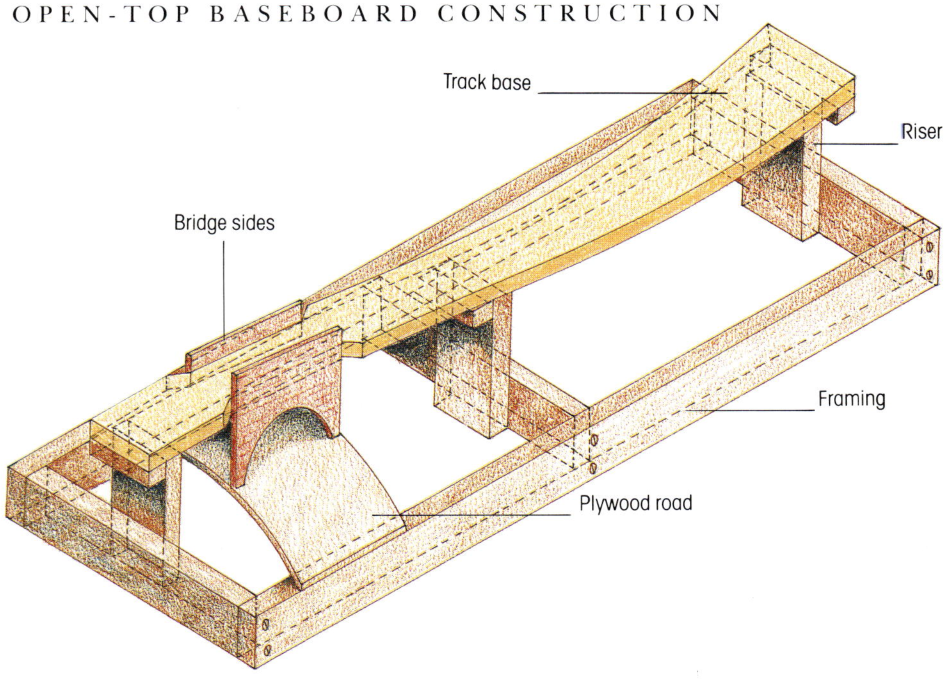

The wood frame is made from 2 in × 1 in (50 mm × 25 mm) timber. By getting rid of the solid top and substituting a narrow track base, supported on risers, greater scope is provided for scenic effects. The diagram shows how an underbridge for road or river can be arranged.

ABOVE

Diesel line-up on an HO gauge layout of Allegheny Midland Railroad. From left to right we have an SP38-2, Alco C-430 and an Alco C-420. A caboose stands on the extreme left. A row of well-detailed buildings forms a pleasing backing to the railroad, but the massed foliage of a densely wooded slope makes a superb backdrop to the model.

A simple rectangular frame is not a stable structure, as it can twist and in so doing places considerable strain on the corner joints. Even where careful workmanship has ensured maximum strength, deterioration can take place in time, particularly on open-top frames where there is no top surface to give additional strength to the unit. The solution is to reinforce the corners. Proprietary metal brackets are sold for this purpose, and they can be secured to the timber with countersunk screws. A slightly cheaper approach is to make a plywood gusset and to pin and glue it under the corners.

With this type of construction, the size of each module should be kept within reasonable limits since each section will, at some time in its life, have to be manoeuvred through the passageways and doors of the home – and probably by the modeller himself. As mentioned before, a 1.2 × 0.6m (4 × 2ft) unit is about as large as can be comfortably managed, and the fact that most DIY stores stock sheets of plywood and chipboard to this size is an added advantage.

It is not essential that all baseboard modules be identical in size, nor need they be rectangular. This is merely the most straightforward form of construction. It is, however, a good idea to make them in mirror image pairs, so that the two units can be stored face-to-face, separated by a pair of end boards. Even if the layout is to be permanently erected, it is still a good idea to do this, since it makes any later move of home far less fraught. It is possible to hinge baseboard sections in pairs to fold into a convenient 'crate' for transit and storage. In this case, the maximum recommended length is 1m (3ft 3in), since it is vital on a portable layout to keep the total weight of each unit down to reasonable limits for ease of handling.

CROSSING PASSAGEWAYS

If a layout is permanent it is often necessary to have baseboards across a passageway. There are three ways of dealing with this.

The simplest arrangement is the 'duck under'. Here, you merely provide a clear passageway under the baseboards so that you and your friends can crawl through. To make this easier, the baseboards need to be at least 1m (3ft 3in) high, the framing should be fairly substantial and padded with foam plastic (everyone is likely to hit it from underneath at sometime or other), and a carpet and a pair of handrails will make transit more comfortable.

Another method is to have a hinged section, which is similar to a bar flap, but the hinges must be raised above track level. The construction has to be fairly strong and provision must be made – usually by means of a hook and eyelet – to secure the section in the upright position.

The third approach is to arrange for the section to lift out completely. This gives greater scope for scenic effects – a lift-out section is ideal for a short viaduct, for example – and the section can be stored on runners under the baseboard when it is not in use.

Weight considerations have led to the development of ply framing. A popular arrangement is to make the rear frame to a depth of 300mm (1ft), so that the backscene forms an integral part of the structure. The use of 6mm (¼in) thick plywood for the framing, with 20mm (¾in) square timber for corner joints, can reduce the weight by up to 50 per cent of a conventional timber frame, while the fact that plywood does not warp is an added virtue. The main problem is cutting a larger sheet of ply into relatively narrow strips. The ideal tool is, of course, a sawbench, which is not a very common fitting. Hand-held power saws, especially if fitted with a guide, are quite effective, while the job can be done with a tenon saw, which is a tedious rather than difficult task. A simpler solution is to plan the job beforehand and have the timber merchant cut the sheet into the required widths on his sawbench. Although this may cost a little and will certainly require advanced warning, as it will need to be done during a slack half hour, it will not only save a great deal of work on your part but will make it a lot easier to bring the materials home in the family car.

Baseboard sections require some form of support. The most obvious is the common trestle, but this takes up a good deal of timber and can work out quite expensive, although the idea of using a small number of trestles to support longer bearers, on which the baseboards can rest, has some merits. Folding legs can be screwed to the baseboards, or it is also possible to provide plug-in legs. On permanent installations, the legs can be simply bolted to the framing.

An elegant solution is to use screw-in legs of the kind sold for DIY furniture projects. Those found in most stores are on the short side, which means that the baseboards tend to be low. This is not necessarily a disadvantage, since this makes them more stable. However, longer legs can be obtained to order, although this is sometimes not easy to do in larger superstores. Smaller hardware stores are more likely to be able to obtain specific items like this.

The main limitation of the sectional baseboard is that complex, multi-level schemes tend to get a trifle difficult to arrange. For this type of scheme, a permanent site, coupled with a skeleton frame, is preferable. The ideal form of construction involves the L-girder principle, which, although complex to explain, is very easy to follow from the diagram on page 35.

Basically, the main framing is supported on pairs of wooden girders with an inverted L-cross section. These are built up from two pieces of timber, glued together. The original specification in *Model Railroader* suggested that screws should be used to hold the two parts together while the glue was setting, being subsequently removed for re-use. This involves a good deal of work, and oval nails do the job admirably, as long as the two sections are initially held in place by means of G-clamps. The suggested dimensions will span about 3m (9ft) between legs. For larger runs, the depth, but not the thickness of the vertical member, should be increased from 75mm (3in) to 100mm (4in).

Joists are laid across the L-girders, secured from beneath with screws. The spacing is not critical, but the distance between joists must be such that it is possible not only to get a screwdriver between them, but also to use it effectively. A space of about 300mm (12in) is about right, but if necessary it could come down to 200mm (8in). It is advisable not to exceed a maximum space of around 450mm (1ft 6in).

Vertical risers are screwed to the joists. These carry cleats at the top, which can be glued and nailed to the risers to economize on screws. The cleats are screwed to the track bases from below. The idea is that all securing screws should be accessible from under

BASIC L GIRDER CONSTRUCTION

For permanent layouts, L girder construction is growing in popularity. With this arrangement, the main supports are inverted L timber beams, with cross joists secured at intervals with screws. Risers carry the track bases. The track bases can be sprung to the required gradient, though the amount of curvature tends to be very slight.

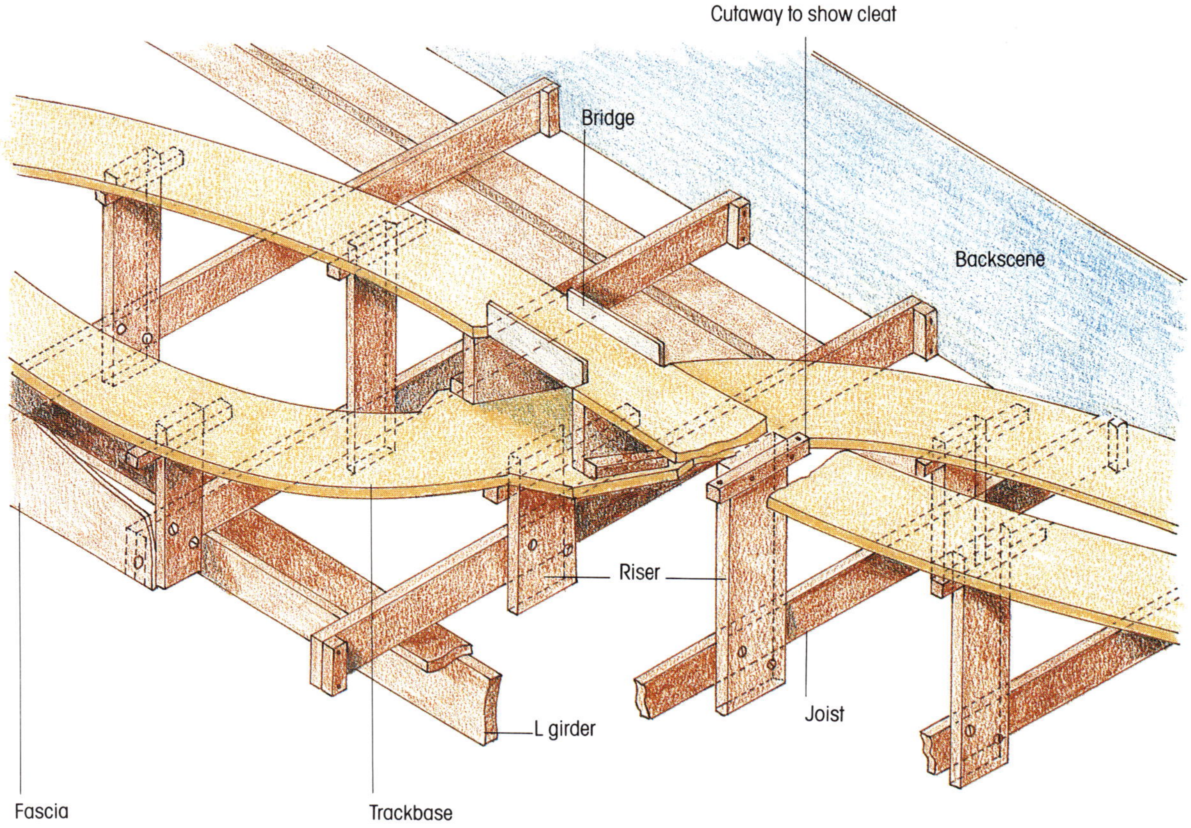

the framing, to make it easy to carry out alterations that may become necessary if it is found that the supports foul some essential part of the layout. The result is a very flexible, open-work framing, which can be reduced to its component parts in the event of a move.

The ease of adjustment makes it a great deal easier to arrange gradients. The risers are held in place with G-clamps while the track bases are aligned, the fixing screws being inserted only when the levels have been carefully checked and found to be in order. An alternative arrangement is to have a single long slot used with a couple of close-fitting screw holes to allow the riser to be held by one screw initially while very fine adjustments are made until the second fixing screw can be inserted.

One of the beauties of L-girder construction is that no specific section of timber is required, and second-hand timber can, therefore, be pressed into use. If you have access to a sawbench, second-hand floorboards can be rapidly cut into strips to form the L-girders and joists, while the scrapbox will provide plenty of material for risers and cleats. Another advantage is that very few legs are needed; indeed, if it is possible to screw things to the walls, the entire layout can be carried on girders spanning the space. This effects a considerable saving not only on materials but on obstructions, which is all the more important since a good deal of work has to be carried out under the framing.

With this type of construction, groups of turnouts are mounted on a large trackbase that has been shaped to conform to the track layout. Fixing the turnout motors and the main wiring can be carried out on this section in the workshop, before the completed units are located on the framing and further trackbases cut to accommodate the connecting tracks. This enables the trackwork to be dismantled and re-erected in a slightly different alignment should a new railroad room be different in size and shape.

CHAPTER SIX

LAYING THE TRACK

Although it is possible to make your own track, it is cheaper to buy ready-made flexible track in metre or yard lengths for less than the cost of components. Some saving can be effected when building turnouts, but these are not the easiest of modelling projects to tackle. No matter how little it costs you to make, a turnout that derails one set of wheels in 10 is very expensive, since the chances of getting a train to run through it are nil. Most people opt for ready-assembled track.

Before we go any further, it will be as well to define our terms. A section of track consists of two rails, which are supported and fastened to cross sleepers (ties). The distance between the inner faces of the rails at the top is the gauge. In model practice, the rail height is given by a code number, which is its height in thousandths of an inch. Sections of track are connected by rail joiners.

The place where one track splits into two for junctions, loops and sidings is called variously a point, a switch and a turnout – and this latter term will be used throughout.

THE COMPONENTS OF A TURNOUT

A right-hand turnout showing the various parts of the unit.

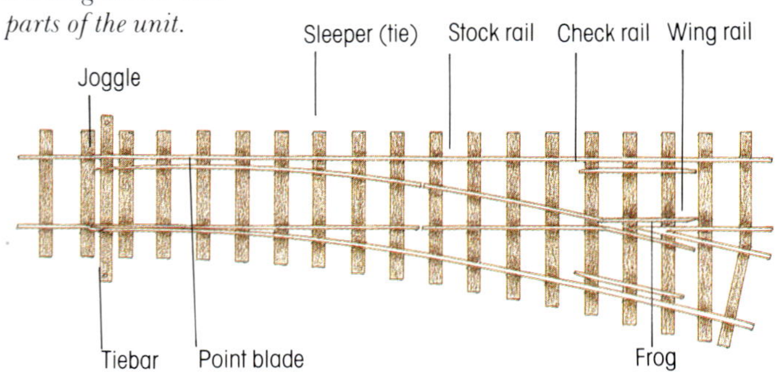

For our purposes, we require track laid to either 16.5mm gauge for HO or OO and 9mm gauge for N. In each case there is an accepted standard for the basic track, for sleeper (tie) spacing and lengths and for rail height. In HO/OO the old standards used Code 100 rail, which could accommodate the deep flanges on certain commercial models. In the US, where wheel profiles, with a few short-lived exceptions, conform to NMRA (National Model Railroad Association) standards, finer section track, using Code 70 rail, has long been available. Recently in Europe, similar track has come on to the market. In N, there has long been general agreement on wheel profiles.

It is possible to use either sectional track with fixed-radius curves and rigid track units, or flexible track, which offers, in theory at least, greater flexibility. However, the rigid sectional track is easier to lift and re-lay to a different formation, and so, paradoxically, it can offer a more flexible layout design. Furthermore, when you are working to tight radii, the rigid sectional track is to be preferred since it is precision made and, when new, is within very close limits of a true arc. Flexible track, on the other hand, even when curved around a precision template, can vary considerably from true. However, as the rail section and sleeper (tie) profile is identical, sectional and flexible tracks may be mixed on one layout.

Whether sectional or flexible track is used, the rails have to be linked together. This is done by means of rail joiners, which are sometimes termed fish-plates, even though they are substantially different in design from the prototype fitting of this name, which is used to bolt sections of full-

FACING PAGE
Intersecting tracks add to the busy nature of this corner of the *Model Railroader* magazine's HO club layout, the Milwaukee, Racine & Troy. Crossing the picture is a train headed by a BQ23-7, with a Wisconsin Central SD45 stationary (one hopes!) at the rear. Note how the railroad-associated structures hug the tracks, filling the space between the lines with activity, whilst the structures to the rear, which suggest the existence of a large town, are modelled in low relief.

neb+w
BERKSHIRE LINES
Milwaukee, Racine & Troy
ART

TURNOUT FORMATIONS

Even the most elaborate track formation is built from standard turnouts and crossings. The diagram shows a number of formations and track units so assembled. Note how the ***crossover****, linking two parallel tracks, is made from two turnouts, back to back. The* ***scissors crossover****, by contrast, allows trains to move from track to track in either direction, and is made from two turnouts and a small diamond crossing. Two turnouts and one large crossing make a* ***double junction****. The* ***double slip*** *is a special type of crossing which permits trains to 'cut a corner'.*

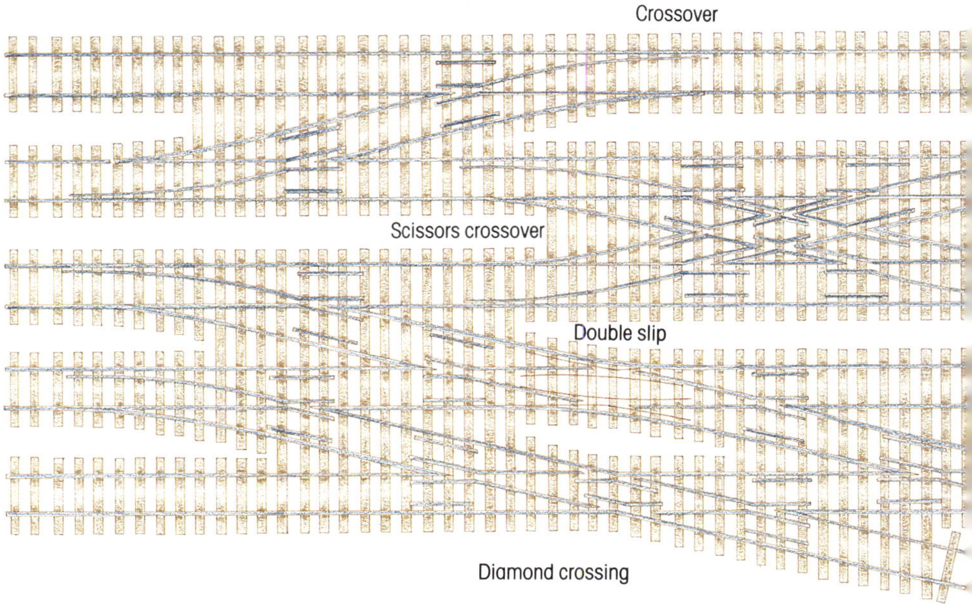

RIGHT

Commercial ready-to-lay track at its best. Peco Streamline OO gauge track and turnouts were used to create this evocative steam-age station. The design, a busy junction with avoiding loops on the main line, allows for intensive operation of long trains.

ABOVE

The sweeping curves of this intricate trackwork show what can be achieved in N gauge using commercial track and turnouts – in this instance, Peco Streamline flexible track.

sized rail together. A model rail joiner consists of a flattened C-section metal strip that slides over the bottom flange of the rail. Insulated rail joiners, moulded in plastic, are provided to enable the tracks to be sectionalized electrically, as is described in the following chapter.

Because rail joiners are considerably longer in proportion than the prototype fish-plate, they occupy rather more than the standard spacing between the sleepers (ties). There are two ways of getting around this difficulty. The preferred method is to cut away the rail fastenings on each end sleeper (tie), allowing the rail joiner to slide between. The lazy way is to leave a gap and slip a spare sleeper (tie) in later. The only fault with this latter approach is that it rarely gets done, and the overall realism of the model suffers.

Some problems may arise if differing rail sections have to be joined, and many people believe it is impossible to do this successfully. This is not so. On the prototype, a special joggled fish-plate is used to ensure that the rail heads line up. On the model, where the rail joiner slips over the bottom flange of the rail, the solution is simpler – the head of the higher rail is just filed away for a short distance to ensure that the wheels pass smoothly across the joins.

LEFT

Candleford Mill depicts an imaginary English country town, with thatched roof cottages facing the station. The track is Peco Streamline, and the layout has a wide range of British ready-to-run locomotives on view.

ABOVE

This OO gauge model shows just how effective standard commercial ready-to-lay track can be made to look. A combination of wide curves, a prototypical track layout and careful attention to the finer details of ballasting create the illusion of reality.

Whether you use sectional or flexible track, the actual laying appears to be the simplest of operations – you link the sections together and fix them to the baseboard. This is, in fact, a fallacy, and it leads to a good deal of indifferent running because it is important that the tracks and turnouts be accurately aligned if the train is to have a smooth passage throughout.

The correct procedure is to begin with the turnouts, for these are, by their nature, fixed units, and you cannot juggle with their size or alignment. On all but the most elementary of layouts, turnouts come in bunches. These formations should be assembled first and the alignment checked carefully by squinting along the top surface of the track. This will show whether the units are properly lined up or if you have introduced a dog-leg or two. Dog-legs – that is, sharp twists in the track where the rails meet at an angle – are a prime cause of derailments. Once you have the formation in place and secured to the baseboard, you can begin to put the approach tracks in position. You will probably find you have to set a second turnout formation some distance down the line, and this, too, must be properly lined up before you can finalize the position of the approach tracks.

The beginner will probably find this alignment a little easier to achieve with sectional track, for the rails of flexible track have to be cut to fit. If flexible track is used, the rails should be cut with a fine-toothed saw, after the exact position for the cut has been firmly marked on the head of the rail. A small triangular file is ideal for this purpose. Cutting through rails is not quite as simple as it looks, for the inner web is very narrow, and the saw blade can catch on this if the cut is forced. A very fine-toothed razor saw will be less likely to do this than the coarser, pin-ended, metal cutting saw, but the latter is a more substantial tool and also cuts faster. It is a good idea to file the base of the cut rail to remove any burr and to ease its entry into the rail joiner.

CURVING TRACK AROUND A TEMPLATE

Although flexible track is easy to curve, for good running this needs to be done accurately. To avoid introducing kinks, it is best curved around a template cut to the desired radius.

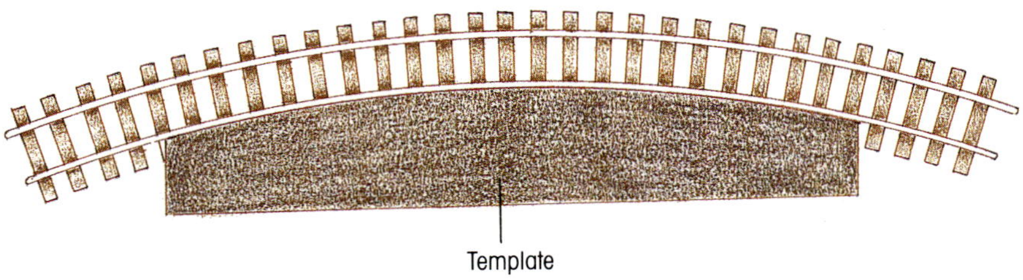

Although the track can be secured through the sleepers (ties) directly to the baseboard, many people like to use an underlay. Commercial foam underlay, slotted to take standard sleepers (ties), is available and gives a fair impression of ballast. However, if the track is pinned down through this material there is a considerable risk of locally compressing the ballast and thus producing an uneven track, which is another good way of getting poor running. If this type of underlay is used, the track should be secured by running some PVA adhesive along the underlay and pushing the track into the slots. The underlay can then be either glued to the baseboard or fixed in place with double-sided self-adhesive tape.

Thin cork sheets can also make an excellent underlay, but in many respects the best arrangement is to lay a sub-base made of insulation board between the track and the baseboard proper. If the sides of this sub-base are bevelled, as shown in the cross-section, you will have created a close replica of the prototype's own sub-base.

The most realistic form of ballast is small stone chippings, which are a near-to-scale replica, in both shape and material, of the prototype. These are available not only from modelshops, but may also be obtained, in bulk, from quarries, where stone dust is usually a nuisance, and a large sackful can often be acquired quite cheaply.

The best way of laying ballast on the model is to put it in place dry, carefully neaten the edges and tidy around the track until it looks right and then use either an eye-dropper or a small spoon to apply a mixture of 60 per cent PVA adhesive, 40 per cent water and two drops of washing-up liquid to the ballast until it is completely soaked. Leave it to dry overnight, and 90 per cent of the ballast will be firmly fixed. Take care not to clog the tie-bars on the turnouts while this is being done.

If the layout has been built on sectional baseboards, special provision has to be made at the baseboard joints. Whether flexible or sectional tracks are concerned, the method is the same. The tracks are initially laid across

CROSS-SECTION OF TRACK

For realism (as well as a degree of sound-muffling), it is a good idea to lay track on a resilient base. Cork sheet, 1/8 in (3 mm) thick is ideal. Loose ballast is then laid on top and fixed with diluted PVA adhesive. Commercial foam plastic can also be used as ballast.

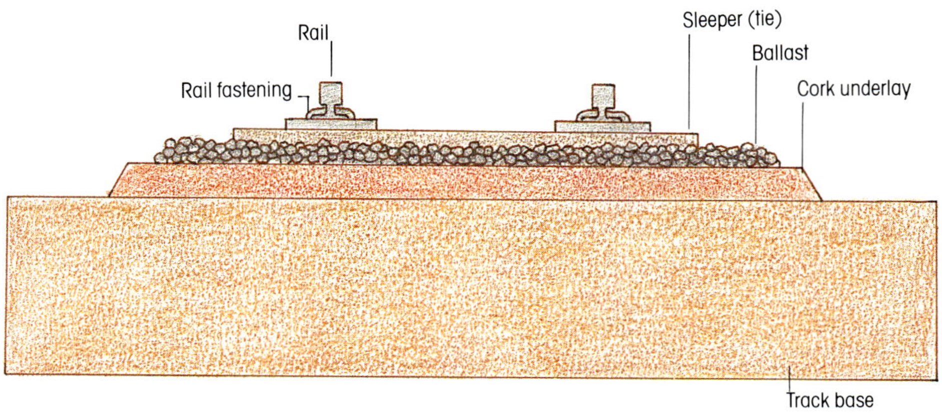

RIGHT

A corner of one of the Fleischmann HO gauge exhibition layouts, showing their flexible track effectively used to create wide curves on this corner station. Deutsche Bundesbahn stock from the same manufacturer is very much in evidence.

CARRYING TRACKS ACROSS BASEBOARD JOINTS

The best way to take tracks across a joint in the baseboard is to lay the track over the gap, solder the rails to small screws driven into the baseboard, and then cut through the gap with a fine-toothed saw.

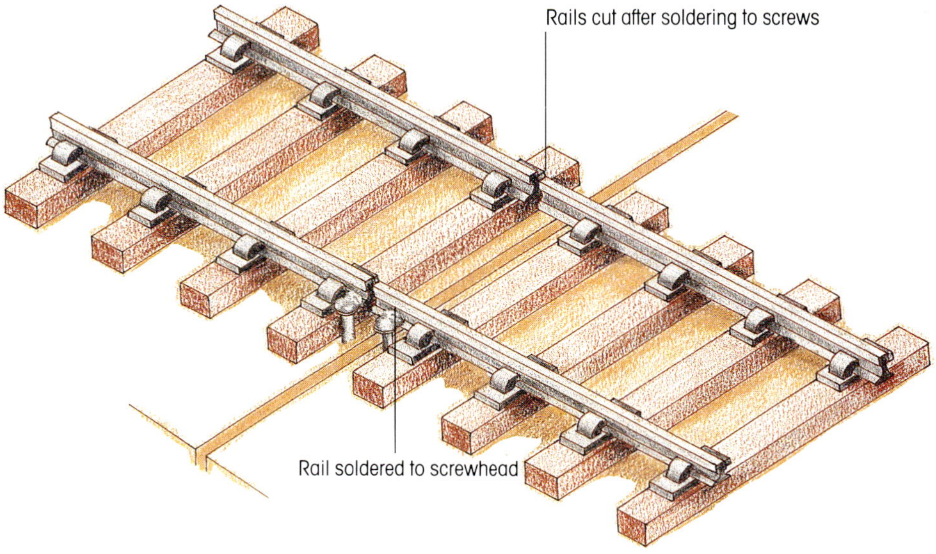

TURNOUT OPERATION BY SOLENOID MOTOR

A popular method of turnout operation involves a commercial double solenoid motor. This is usually mounted below the baseboard, with an operating crank above, and linked to the tiebar with a length of wire. All manufacturers provide detailed mounting instructions with each motor.

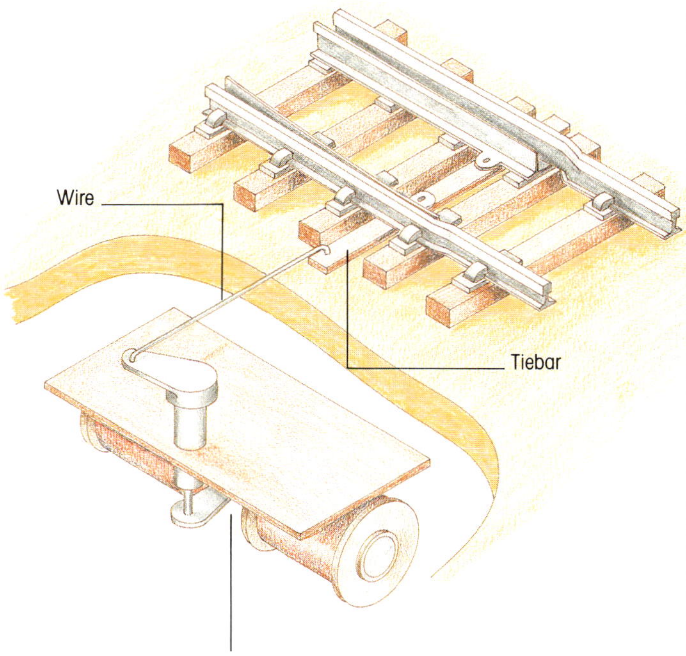

TURNOUT OPERATIONS

There are many ways of operating turnouts, the simplest of which is to use the mechanism on the turnout itself. This has the virtues of simplicity and reliability, and it falls down only when the turnouts are out of convenient reach. It is, however, a little basic, and something more sophisticated may be preferred.

At the other extreme is the electric turnout motor. The commonest form incorporates two solenoids and is operated by a single spurt of current through the appropriate solenoid, which moves the armature across and, through a simple linkage, shifts the turnout blades. There are several designs on the market. Fitting and wiring instructions are provided with each unit, and, in most cases, the user has the option of mounting the motor above or below the baseboard. The main exception to this is with the sectional track produced by the larger manufacturers, on most of which the motor is mounted unobtrusively alongside the turnout.

Turnout motors are, of necessity, costly. They are invaluable for remote locations, but, for the most turnouts, which will be found close to the control position, a simple mechanical system can be arranged.

A popular alternative is the wire-in-tube. Here, a thin steel wire is threaded through a length of small-bore copper tube. The action is similar to the sort of cable that is used for manual bicycle brakes and gear shifts, except that the outer tubing is rigid. It can be curved to 50mm (2in) radius, and sharper turns can be made by means of bell cranks. Coupled to a mechanical lever frame, this system will work turnouts up to about 2m (6ft) away, depending on the degree of curvature imposed on the tube.

An even simpler and cheaper system is the sub-baseboard push rod. A stiff wire runs under the baseboard. The wire terminates in a simple knob (a section of ball-point pen body secured to the rod with epoxy resin is more than adequate for this), while a thinner stiff wire is wound around the main wire and passed through a 6mm (¼in) diameter hole in the baseboard surface beneath the hole in the turnout tie-rod. The other end of the wire is wound around the main rod and secured with epoxy resin. The coat-hangers supplied by dry-cleaning firms when returning garments are an ideal source for the main wire.

TURNOUT OPERATION BY PULL-ROD UNDER BASEBOARD

Simple and cheap, the pull-rod system allows the turnout to be worked from the baseboard edge. The main requirements are a good length of stiff wire, a short length of springy wire, an improvised knob (a slice off a cheap ballpoint pen will do), and a wood block drilled to take the far end of the stiff wire.

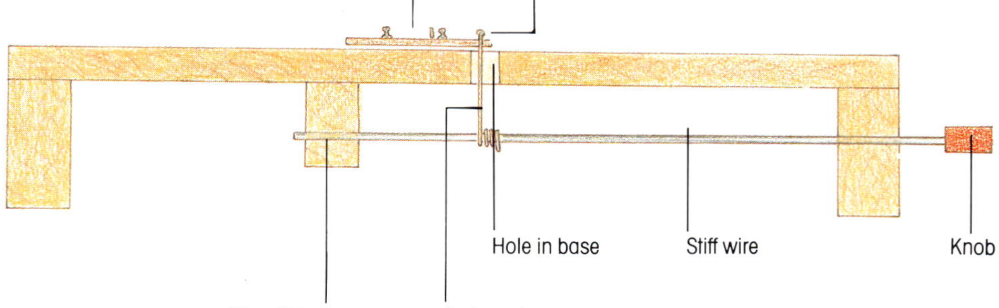

RIGHT
Lineside industry surrounds this section of an O gauge layout of Maidenhaiste, UK. The buildings have been weathered effectively, and it can be seen that some planks have been recently replaced.

the joint and secured in place in the usual manner. Four small wood-screws should then be driven through the top surface and into the framing next to the rails, which are then soldered to the head of the screws. Once this is done, the rails can be sawn through above the gap. In this way they will be correctly aligned. Do not place turnouts across a baseboard joint for obvious reasons.

Track laying with ready-made track is really little more than a matter of taking pains to see that the tracks are correctly aligned in the first instance and that nothing is distorted or misplaced in the process.

TURNOUT OPERATION BY WIRE IN TUBE

Wire-in-tube operation provides a simple link between a manual lever frame and a turnout. Similar in principle to brake cables, fine piano wire is threaded through a small-bore copper tube. This must be done before the tube is bent.

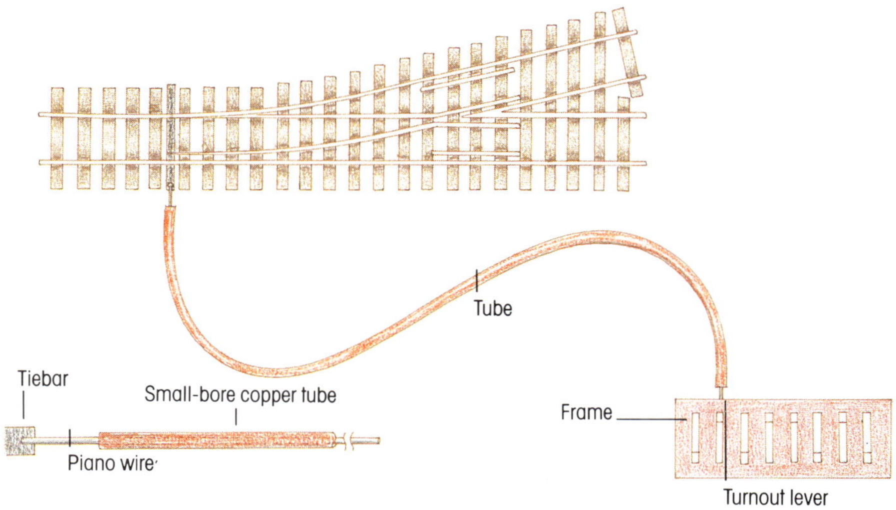

CHAPTER SEVEN

WIRING AND CONTROL

Although there is no real limit to the degree of sophistication that may be introduced into model railroad control, it is as well to begin with simple yet effective methods that do not require a doctorate in electronic engineering to comprehend. The main requirements are a power unit/controller (see page 47), some low-voltage wire – single-cored bell wire is suitable – some switches, a soldering iron and a little commonsense.

The following rules explain what has to be done to wire a layout.

1 Take two wires from the power unit to the length of track from which the turnouts radiate. On most layouts several feed points will be needed.

2 Label one wire 'feed' and the other 'return'. Section switches must be in the feed wire; it is advisable to take all feeds independently to the control position but all returns may be connected together. The feed and return wires and rails must not come into contact with each other.

3 You must put an insulated rail joiner in each rail where turnouts are back to back. There must also be at least one pair of insulated rail joiners in any continuous circuit.

4 To isolate small sections – in locomotive sidings or platform roads, for example – place an insulated rail joiner in the feed rail. Bridge the electrical gap thus formed by means of a pair of wires leading to an on-off switch at the control position.

BASIC TWO-RAIL WIRING

This diagram shows how a simple oval layout should be wired. Note that each feed has an off/on switch so that the section can be cut out, allowing a train to be held stationary whilst another one is moving.

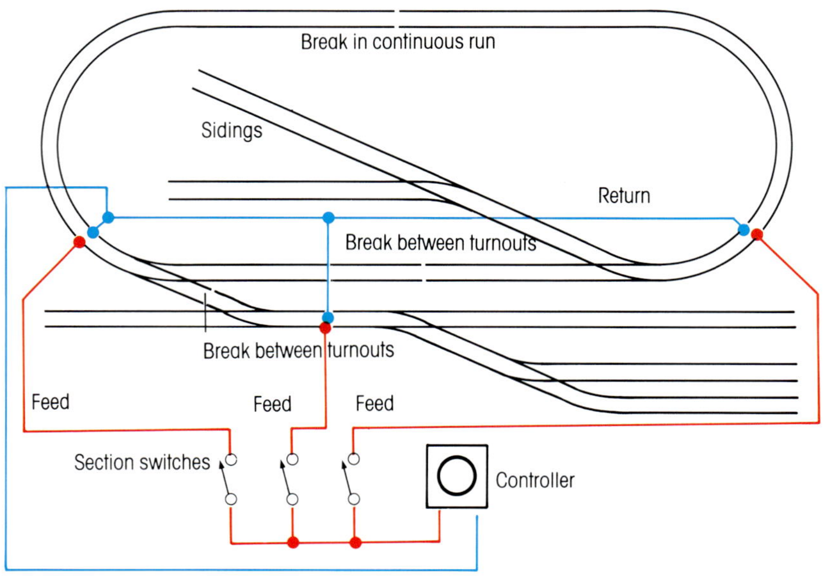

FACING PAGE
Inspired scenic modelling in a corner of a Deutsche Bundesbahn HO gauge layout. Although the individual models are all of commercial origin, and completely standardized, they have been combined with skill to create a wholly satisfying and individual scene.

It is necessary to be able to isolate locomotives at various places on the layout – usually in loco sidings or at the ends of terminal roads. This is done by isolating one rail and bridging the gap with an on/off switch.

WIRING ISOLATED SECTIONS

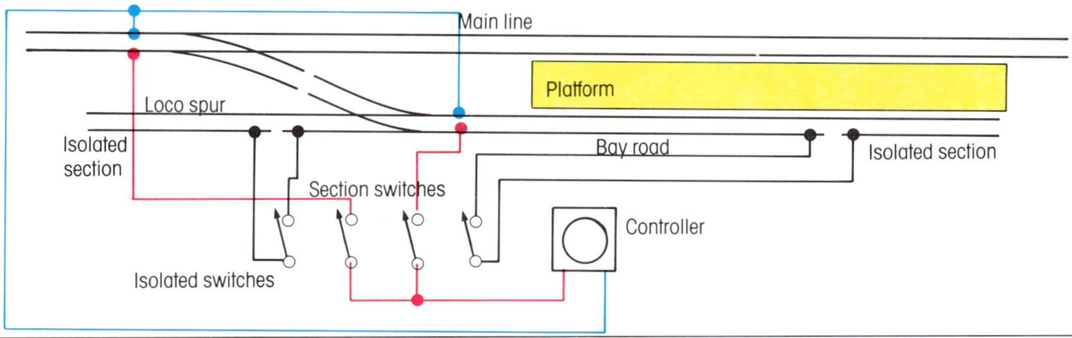

These rules cover the basic requirements for a simple layout on which one train can be operated, with other locomotives held isolated in sidings, loops and dead ends.

To run two trains independently, you do not need electrical gimmicks; rather, you need two independent stretches of track and two independent controllers. A common example is a double-track layout, where the trains normally circulate in opposite directions. Two-train operation depends more on layout design than on electrical refinements, as the prime requirement is that it should be possible to move either train at will without it running into the other.

If reverse loops or triangles are introduced into the layout, the feed rail will meet the return rail at some point in the circuit, with unfortunate results – the integral cut-out in the power unit will operate and shut off the traction supply so that nothing will be able to run.

There are two straightforward solutions. One is to isolate completely both ends of the loop and to feed the isolated section through a reverse switch. If, as is usual, the reverse switch is fed from the main controller, the

BELOW

A small control panel built into the baseboard of a 4mm scale layout. The left-hand end incorporates a commercial electronic controller module, while the main switches below the miniature track diagram began life as part of an office telephone exchange. A lot of the older telecommunications equipment has been used to good effect on model railroads, the two systems use similar circuitry.

SOLDERING

Most model railroad electrical circuits, apart from those based on train set layouts, rely on soldered joints. All joints made by means of crimping or twisting wires together are suspect, but a properly soldered joint will conduct current indefinitely. If you have a mains-powered soldering iron, of 15 or 25 watt capacity, together with a supply of cored solder, the job is quite simple. Begin by plugging in the iron, which should be resting in a stand, and wait for the bit to heat up. Allow at least one minute for this.

While the iron is heating, prepare the joint. The wire and the tag or other part to which it is to be attached must be perfectly clean. A quick scrape with a craft knife or the point of a small screwdriver usually does the trick, but in extreme cases – if you are soldering to a second-hand switch, for example – it may be necessary to polish the surface of the tag with a piece of fine emery paper ('wet and dry' paper).

Place the wire against the tag; if the tag has a hole in it, loop the wire through this. Make sure that the tip of the iron is clean; it is generally necessary only to wipe it over with a piece of damp sponge, but if it has got badly oxidized, rub it with an old file. Apply a small amount of solder to the iron to tin the bit, place the hot iron against the joint and apply the cored solder. The solder should fizz as it and the inner flux melt, and a small blob of silvery molten metal should flow round the wire and on to the tag. It is vital that the molten solder is mirror-bright and flows freely so that it amalgamates with the clean metal surfaces.

If the solder merely turns a dull grey and moves about like putty, the iron is not hot enough and the result is a 'dry joint', which not only lacks strength but has a high electrical resistance, the last thing we want. However, as long as the iron has been switched on for at least one minute, it should be hot enough to enable a proper joint to be made. As a final check, tug the wire smartly. If the joint is properly made it will hold, while a dry joint will almost invariably part. When in doubt, apply the iron again, and watch the solder melt and flow, adding a little extra for good measure.

The solder will fail to flow on to either the wire or the tag if the metal was not properly cleaned. Another reason could be that one part is made of aluminium, a metal that will not solder unless special solder and fluxes are used. This is, however, very unlikely unless you decide to solder a wire to a metal can.

Wires should also be soldered to rails. It is essential to scrape the web and base of the rail with the point of a small screwdriver to remove any coating of oxide. It is also as well to work quickly. If you leave the iron in contact too long, the plastic sleepers will start to melt, and for this reason, too, it is inadvisable to solder directly to a point.

train must stop on the loop so that the polarities can be reversed. However, if the reverse loop is fed from another controller, the main control can be reversed while the train is travelling around the loop.

As triangular junctions are almost always provided on model railroads so that a train from a terminus can run in either direction on to a continuous run and vice versa, the two-controller method is very convenient, with one controller for the terminus and the other for the main-line loop. Both controllers have to be set in unison to allow a train to pass across the gaps.

These simple, inexpensive arrangements are sufficient to provide adequate control on the great majority of model railroads.

TRIANGLE WIRING

As in the previous diagram, triangular junctions also bring opposing rails together. As this formation, on a model, is usually employed to link a terminus to a continuous run, a very simple solution is to feed the terminus road from a completely separate controller. And although the latter and the main-line controllers need to be set in unison, in practice this is automatic.

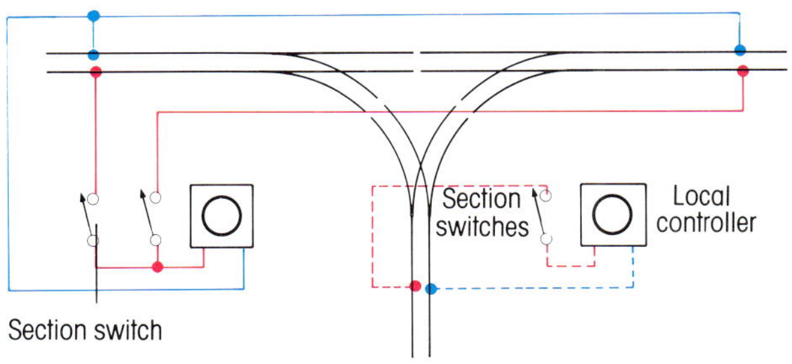

POWER SUPPLIES

Model railroads can be run from batteries, but this is an expensive option. The best arrangement for the newcomer is to use a combined power unit/controller, a sealed device that plugs into the mains and delivers controlled and uncontrolled low-voltage supplies to the layout. With one small proviso, it is a safe and reliable method.

The snag is that most power units are fitted with a short mains lead of 2m (6ft) or less in length. It is highly unlikely that it will reach from the layout to the mains socket, and it will need extending. The only simple and safe way to overcome this is to purchase a ready-made extension lead, fitted with a suitable plug and socket. Choose a lead with at least two socket outlets, although four are more useful as there is usually need for extra supplies for soldering irons, electric drills and so on. Only an experienced electrician should attempt to extend the lead itself.

The lead should be long enough to avoid having it trailing across a walkway. More accidents are caused by people tripping over electric cables than from electrical faults in the cable.

The power unit/controller is provided with a speed control knob. This may be of the centre-off type, in which the knob is turned to either side to effect reverse, or it may have a separate reversing switch, in which case the speed control is over the whole movement of the knob. The supply is nominally at 12 volts dc, but the speed control knob effectively reduces the voltage in one fashion or another. There is no need to ask why, or to understand how the motor revolves, or what makes it go in reverse when the polarity of the feed is reversed. Just accept that it works as required.

Better units also have an uncontrolled 16 volts ac output to operate turnout motors, lighting and so forth. Some units are provided with two control knobs. Although these are superficially attractive, for two-train control it is better to have two separate units, since it is more convenient to have two operators, each controlling one locomotive.

Choose a unit from a reputable manufacturer, as this will ensure that it conforms to the accepted standards of electrical safety. Most model railroad controllers do, in fact, exceed the bare requirements, but as long as the case is not damaged and the covering of the mains lead is not frayed and the plug is not faulty, they are perfectly safe. The only circuits you can touch are the safe low-voltage outputs.

REVERSE-LOOP WIRING

On a reverse loop, the left-hand rail meets the right-hand rail – which, for two-rail wiring, is not a good idea. One solution is to make the entire loop an isolated section which is fed through a DPDT (double-pole/double-throw) switch. This allows the current to be reversed in relation to the main line, whilst the train is standing on the loop.

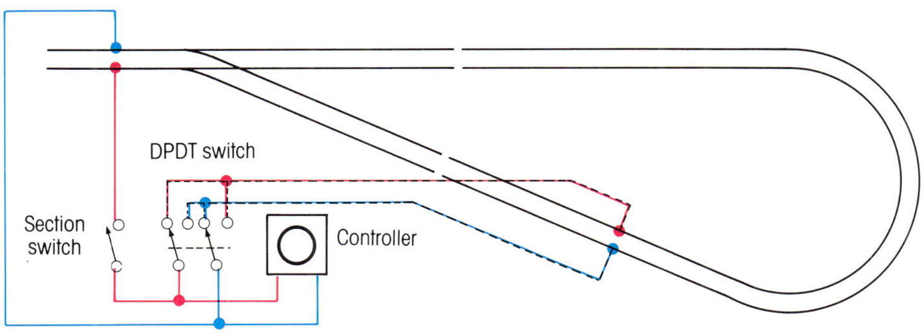

CHAPTER EIGHT

MODELLING THE LANDSCAPE

The terrain of any locality is highly characteristic, not just of a country but of a specific part of that country. Even tiny Liechtenstein has a low-lying, undulating area bordering the Rhine as well as its mountain range. At first sight this might seem to set the railroad modeller a considerable problem. However, in most cases the amount of landscape in the model is limited, and the most that can be done is to represent one of three basic stereotypes – mountain, rolling hills or open plain. Since the difference between the last two is the angle of slope, we are left with two basic forms of construction, rocky outcrops and soil, covered with different types of vegetation. There is, of course, also the seashore, but fundamentally, we are concerned with rocks, shingle or sand, which are only slightly more detailed versions of the two main land forms.

However, before the landscape can be created, it is necessary to create a foundation.

FACING PAGE

This superbly detailed HO gauge 2-6-0 mogul type is typical of a class of locomotives used for general purpose work on US railroads from around 1880 to 1920. The realistic trees in the background are not the product of a modeller's skill, more a credit to the modeller's eye, for they are natural twigs, carefully selected from garden trimmings.

ABOVE

An OO gauge layout in the early stages of construction, showing the preparations for the landscape. A wire-mesh foundation wire is supported on shaped profiles and covered with newspaper, prior to an application of plaster. The foundations for a road and overbridge can be seen to the right – that is, a strip of man-made board, cut to shape and supported on wooden piers.

FAR LEFT AND LEFT

Early and final stages in the construction of tunnel portals on a 2mm-scale layout. Wire mesh on prepared supporting profiles is covered with plaster, then a final coating of scenic scatter material is glued in place. The technique is straightforward, but careful study of full-sized landscape and attention to detail is needed if the effects are to be as realistic as these examples.

F & S Ry

FORMERS FOR LANDSCAPE

Landscape is supported on formers, which can be cut from any convenient scrap material and shaped to the profile of the land. These are glued or pinned into position, and can, if desired, carry the track base. It's not a bad idea to build a small-scale mock-up from card to get a clear idea of how the formers should be cut.

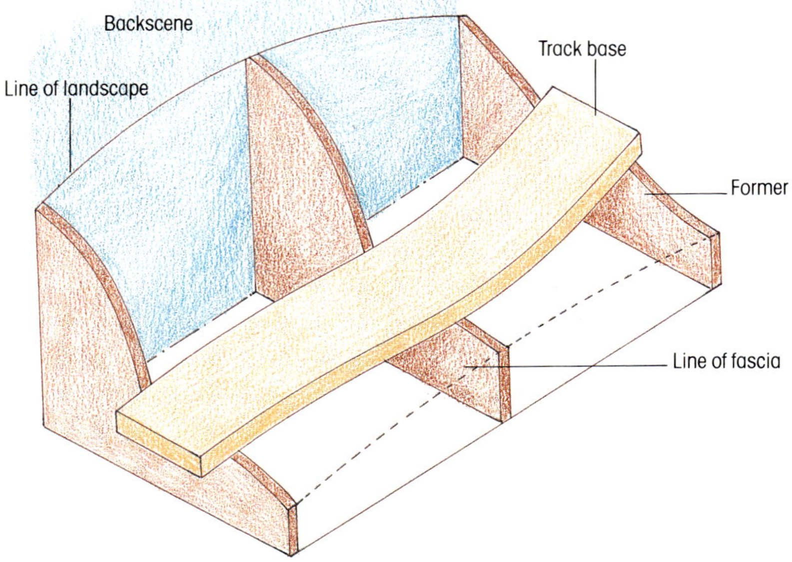

The most popular system involves the use of profiles, which are often miscalled contours. These are vertical forms cut from any convenient material to present the approximate desired cross-section. Very large expanses of land, of the kind found in the 'railroad in the landscape' type of model, require an egg-crate system of cross-profiles. This calls for either a good deal of pre-planning or a lot of trial and error, because the crossing units need to coincide.

When the foundation is in place, it is covered by a sub-base. The most popular material for this is fine-mesh chicken wire, which is most likely to be found in garden centres. This wire is virtually self-supporting over fairly wide areas – up to 300mm (12in) square is perfectly safe – and can be easily teased into the most complicated shapes. It is tedious rather than difficult to trim it to size with a pair of wire cutters, but do remember to wear heavy-duty work gloves, since the wires can easily gash your hands. It can be tacked to the foundation profiles by any means to hand, but a staple gun, which can be wielded in one hand while the wire is held

BUILDING THE LANDSCAPE

Once the formers are in place, they are covered by a fine wire mesh. This is then covered with plaster – when dry, a top coat should be applied and moulded to the desired shape.

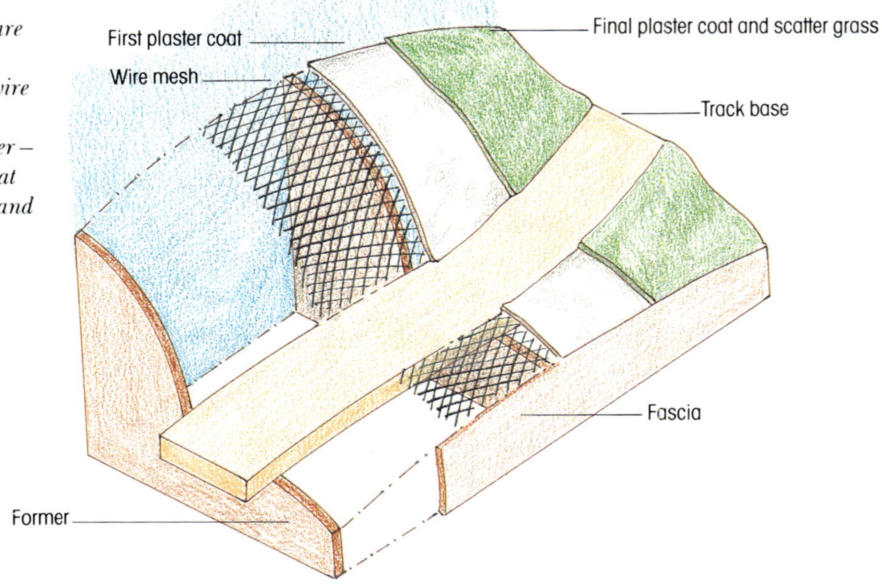

RIGHT

A section of a Z gauge model, built for publicity purposes by Märklin and shown at a recent Nuremberg Toy Fair. The model depicts a spectacular section of the Deutsche Bundesbahn main line, the *Geisingler Steige*.

in place with the other, is definitely the most convenient.

It is just possible to apply common building plaster directly to the chicken wire. If you choose to do this, use a fairly stiff mix. However, it is less messy to apply plaster-soaked newspaper, when the mixture needs to be nearer the consistency of batter. Place small squares of newspaper in an open tray of plaster and apply them to the chicken wire. To ensure long-term durability, it is best to apply two coatings, and then leave the surface to harden overnight.

This is, essentially, a messy process, but good fun for all that. Three simple precautions make it even more enjoyable. Wear old clothes, that is obvious. Cover the floor with a sheet of plastic to catch the inevitable drips, which is equally obvious but frequently forgotten until too late. Last of all, have a bucket of piping hot water and an old towel to hand to rinse the muck off your hands whenever you feel the need. It will cool down as the work progresses, but unless the area to be covered is very large you will run out of landscape before the water gets really cold.

BELOW

Landscape under construction on a 2mm-scale layout.

BELOW

The early stages in landscape construction. Wire mesh is spread over a supporting framework and moulded to the required shape. The road and overbridge base is cut from a piece of board and supported on wooden piers; later, these will be covered with the modelled abutments and bridge girders.

While you cover the wire it will also be necessary to protect the track by covering it with strips of plastic, held down on either side with map pins (small pins with large, brightly coloured, round plastic heads), which are quite easy to remove later on for re-use. When the foundation is hard, it will be apparent that the top surface is covered with a slightly irregular hexagonal pattern, which is caused by the chicken wire. This has to be covered with a top surface of plaster, which is shaped to the required profile. At this stage the technique varies according to the basic type of terrain. Rock faces, paradoxically, are the simplest to arrange. Coat the surface with plaster and then press on to the wet surface a sheet of crumpled aluminium kitchen foil. When the plaster is dry, which should take a couple of hours, peel off the foil to reveal a nice irregular surface, which only needs coating with a fairly dilute paint to create a good effect. The overall appearance is greatly improved by judicious carving when the plaster is hard, but allow some 24 hours to elapse before doing this.

A neat way of producing stratified rocks is to break insulation board into long, narrow strips and to glue them, one above the other, aligning the 'strata' to the required inclination. This is probably the least messy arrangement possible and, when carried out carefully, is remarkably effective.

FACING PAGE

This dramatic shot demonstrates the scope of N gauge, for not only have we a large bridge in a reasonable space, but the locomotive, a 4-6-6-4 Challenger mallet articulated is by no means a small item in itself. Yet the entire structure is brought into proportion by the soaring mountains on the backdrop.

ABOVE

Detail can make all the difference . . . The bridge abutments are being picked out with contrasting colour to emphasize the moulded relief on the formed plastic walls.

BUILDING STRATIFIED ROCKS

Stratified rockfaces can be built up from layers of insulation board, broken from the board to give a rough edge. Any gaps at the bottom can be covered by small 'rocks' or a build-up of 'soil'.

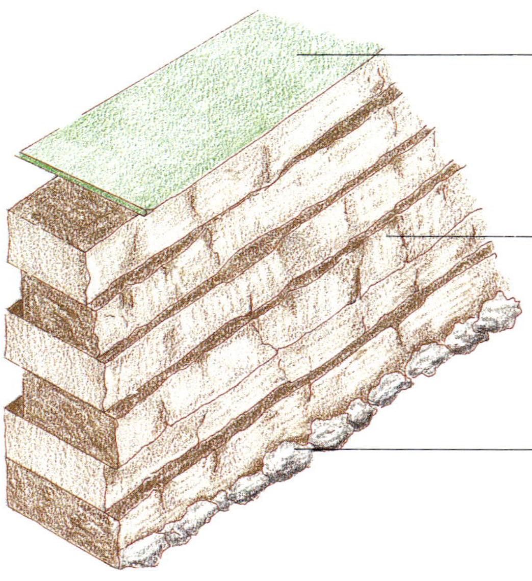

ABOVE

Realistically modelled landscape on a N gauge layout provides the ideal setting for this 1930s steam age scene.

If you wish to reproduce soil cover, the plaster must be smooth and slick so that it flows over the surface realistically. An old, 25mm (1in) wide paintbrush, dipped from time to time in clean water, is the most suitable finishing tool. Undulating grassland can also be reproduced in a variety of ways. One very simple method is to mix coarse sand with the final coat of plaster. When painted, this looks remarkably effective.

As an alternative, one of the finer scenic scatter materials that are available commercially can be used with a plaster mix, although most people prefer to apply these to a glued surface. A suitable base can be prepared by mixing PVA woodworking adhesive with an equal proportion of water and adding a couple of drops of washing-up liquid. This is applied with a brush and the material dropped on to the surface.

Once the main groundwork is complete, the vegetation needs to be added. Trees, hedges and shrubs are obtainable from modelshops, but many modellers prefer to create their own. Hedges can be made from rubberized horsehair, which is used in top-class upholstery, from foam plastic or from plastic pot scourers (the cheaper variety

FACING PAGE

This spectacular alpine scene was lovingly created in a fairly large area, occupying all available space up to the ceiling. Commercial building kits have been carefully located around the landscape to create the atmosphere of the prototype, with smaller scale models placed towards the back to create the illusion of distance by careful use of perspective. The construction of such a model obviously requires many painstaking hours, a large number of kits, as well as a forest of model trees.

LEFT

Scenic artistry at its finest. The bridge abutments and dry stone walls are made from modelling clay, each stone having been individually scribed and shaped; the fences and gates were modelled from strip material; whilst the rock face was scribed in plaster. Scenic scatter materials were used to create the realistic rough grass and undergrowth.

found on market stalls seems superior to the top-quality brand names). Whatever is used, a good deal of teasing is needed to produce the irregular shape and semi-open nature of the full-sized hedgerow. A neatly clipped privet hedge, on the other hand, can be effectively reproduced from a trimmed strip of foam plastic.

Shrubs can be made from teased pot scourer or produced from dyed lichen, which can also be used to create realistic open hedgerows. More difficult are garden vegetables, and most rows of vegetables need to be individually crafted from modelling clay, although rows of beans and peas are not too troublesome once it is realized that the poles must be made from wire if they are not to look thick and ugly. A vineyard calls for a good deal of patience, since its essential nature is given by the serried ranks of vines.

Trees are, however, the scenic modeller's main challenge. There are two possible approaches. One is to hunt through the garden to find suitable twigs; coat the ends with glue and attach dyed lichen. This is relatively simple and produces a tolerable result, although trees made in this way look best when grouped *en masse*.

Alternatively, individual trees can be built from wire. Take a hank of soft, thin wire and twist the lower part into a trunk. Cut the loops short at the other end and tease the wires into a rough tree shape, which can be painted by an aerosol paint. This is a

MAKING TREES FROM WIRE

Trees can be made on a twisted wire base, the 'foliage' being provided by commercial scatter material. With care, convincing models of actual species can be produced.

BELOW

US freight trains tend to be *very* long indeed, something normally difficult to achieve on a model, but this N gauge layout of the Southern Pacific, Mojave subdivision, achieves the near impossible by locating the train on a spiral section of the track. The scenic modelling is extremely effective in depicting the arid nature of the terrain, whilst the rear hills are neatly blended with the painted backscene.

LEFT

Pouring resin into prepared beds to form small ponds – in this instance, part of a sewage farm on a 2mm-scale project.

ABOVE

These trees made from a wire skeleton, with commercial foliage material glued in place, look ready to be inserted into the landscape.

relatively quick process when carried out on a batch basis. When the paint is dry, the branches should be coated with glue and rolled in a heap of scatter foliage, then set aside to dry again, before more squirts from aerosol paints are used to highlight the foliage. The resulting trees should be sorted into three categories. The most realistic are put into prominent positions, often standing alone. The second best are used around the edges of woods and copses, which are built mainly of the third variety; these are reasonably effective when used *en masse*. Providing the modeller has a fair idea of what a tree looks like in the first instance, the end products are remarkably good.

Before we leave the landscape, three other features should be considered – walls and fences, roads and water. Walls can be modelled from moulded or formed plastic sheet representing brick or stone. The most important thing to bear in mind is that the courses of brick and coursed stone walls are always horizontal, while freestone and dry-stone walls follow the sweep of the land. Fences always have vertical posts, at least when they are in good order. Oddly enough, modelling a post-and-rail fence in strict prototypical fashion is nowhere near as difficult as it appears as long as you make a number of spacers beforehand to hold the rails at the required level while the glue is setting.

Roads are simply constructed from shaped ply, hardboard or stiff card and are installed after the profiles are in place. A thin coat of plaster can be laid over the surface to provide a reasonably smooth surface, and it may be shaped to the appropriate camber by dragging a card former across the top surface. The meticulous modeller can even include potholes and patches. A simple way of simulating tarmac is to paint the top surface black and then sprinkle domestic scouring powder on top. Concrete is produced by painting the surface.

Water, in the form of rivers, lakes, canals, ponds and the sea, often forms an important part of a model landscape. The problem is not so much that it is difficult to

ABOVE

Water forms an integral part of most landscape, and here we see how effectively a small lake can be blended into its surroundings.

CREATING PLACID WATER

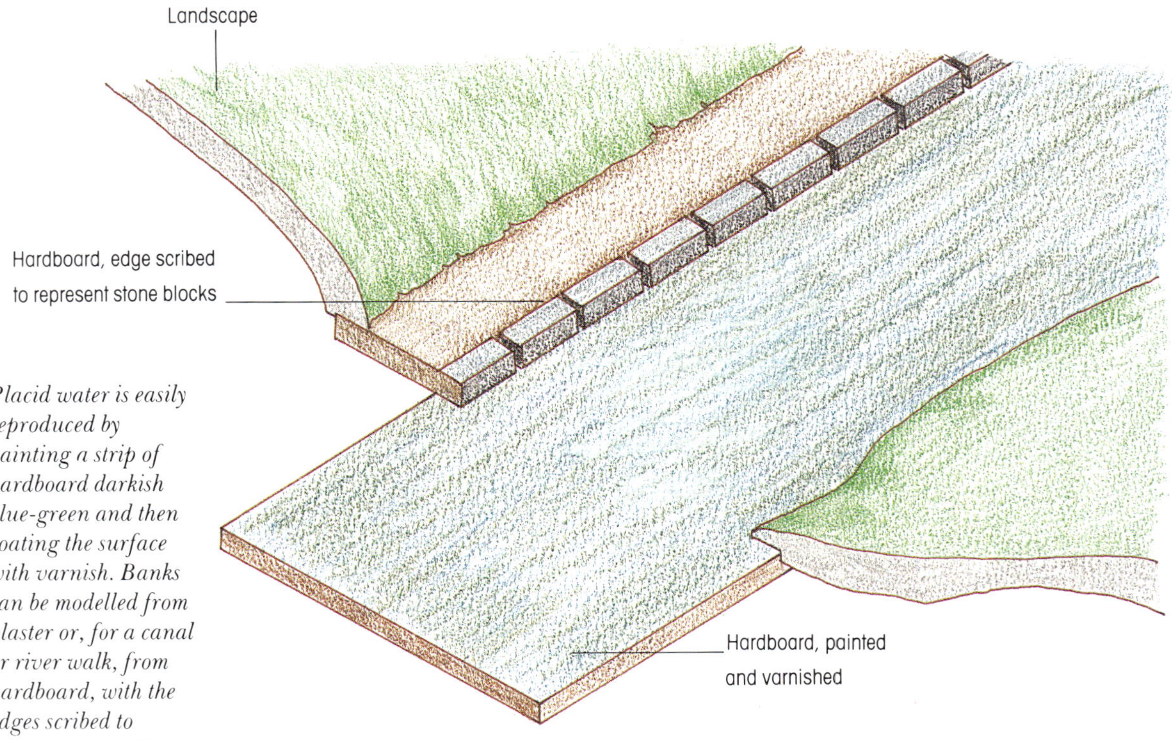

Placid water is easily reproduced by painting a strip of hardboard darkish blue-green and then coating the surface with varnish. Banks can be modelled from plaster or, for a canal or river walk, from hardboard, with the edges scribed to represent stone blocks.

BACKSCENE

The backscene is a simple vertical surface at the rear of the model. It is sometimes placed along the wall (indeed, it can be applied directly to the wall), or it may be placed some little way in front of a series of loops or another part of the model you wish to hide from view. The favoured material is 3mm (⅛in) thick hardboard, which is both readily obtainable and economical in terms of cost.

Commercially printed backscenes are available, and it is often possible to cut up posters or even illustrations from magazines and produce a collage. Indeed, some commercially produced sheets are made with this in mind. However, many of these have one serious fault. They are not, in the strict sense of the word, backscenes, but rather flat pictures of towns and villages. It would be more accurate to say that they show, at best, the middle ground, although in some cases the picture is so detailed that it is arguably including a foreground as well.

Because the backscene should be only a misty representation of reality, it is well within the capability of most modellers to produce their own. No great painting ability is needed, but only a flat surface, a flat brush and a pot of pale blue emulsion paint. Sky is the universal backscene. Indeed, if low-relief models are included, nothing more is strictly necessary.

To get the most from a plain backscene you should introduce some suitable features just in front of the scene itself. This might be no more than a hedge or railroad fence to mark the boundary of the railroad. Hills and even mountains can take the form of cut-out profiles made from hardboard or expanded polystyrene sheet, which are placed in front of the backscene proper. The latter allows for a certain amount of carving to create an additional illusion of depth. (Compare with 'Low-relief Models' on page 78.)

model as that it can be reproduced by many means. For placid, turgid water there is nothing to equal hardboard or ply, which should be painted dark colours and then coated with polyurethane varnish. Another method of simulating level stretches of water is to use crumpled Cellophane, although the main problem today is in locating a source of supply. An alternative method is to model the stream bed, then lay a sheet of plastic over the top, finishing the banks to match the modelled bed. This system can also be used to make lakes and ponds. For added realism, paint the underside of the plastic with streaks of pale blue and green. It is possible to insert small boats, swimmers, cattle and even bridge piers into holes in the plastic sheet. There is no need to make a perfect fit on the item in the water, since the joint can be disguised by a fillet of plastic cement, which can be eased into the gap with the help of a brushful of plastic solvent.

Another way of simulating water is to use two-part transparent casting resin, which is sold for encasing small items in a decorative block. Before you can do this, however, you must ensure the river or pond base is absolutely watertight; more than one modeller has wondered where the resin was going un-

MODELLING A LAKE

A good way of modelling a lake or other stretch of water is to shape the lake bottom in plaster, adding any detail you like. Then, cover it with a sheet of transparent acetate, which can be painted in streaks on the underside, and model the banks to match the bed.

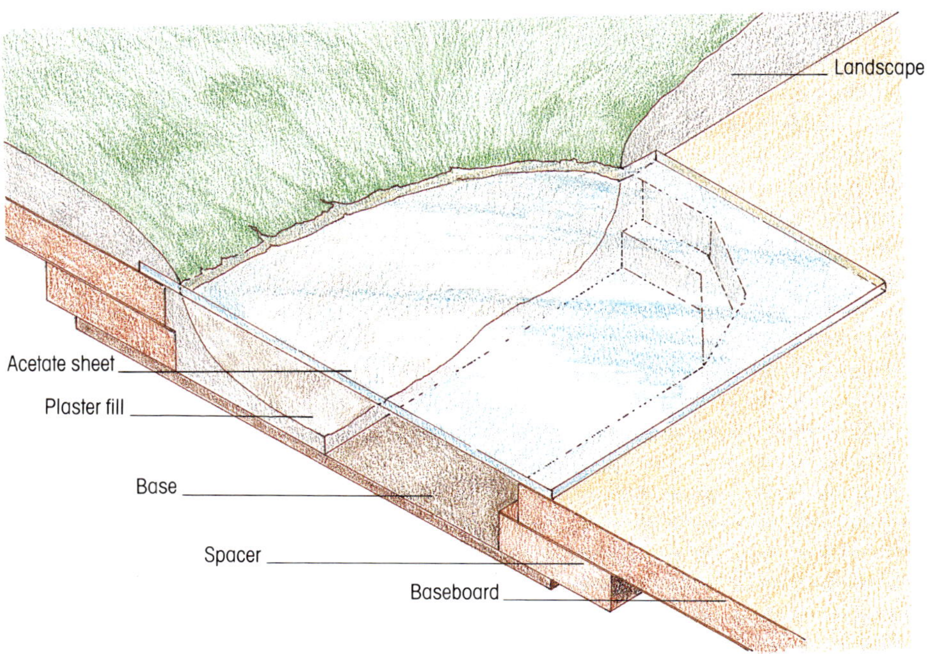

RIGHT

An idyllic British pre-war rural scene on the North Devon Model Railway Club's Penhaven Bridge layout in OO gauge. The train is headed by a Manor class 4-6-0, and the leading vehicle is a Siphon G bogie van, originally built for the carriage of milk in churns. The station is a typical late addition to the line, with wooden platforms supported on square timbers from the embankment sides. The herd of cattle is carefully placed to follow their usual erratic path into a well-known pasture.

til he thought to look underneath the model. A large heap of casting resin does no good at all to a carpet and is not exactly kind to the floorboards either.

So far we have considered relatively level stretches of water. If you want to model a running stream, generally down a mountainside, you must first ascertain that your modelled bed is correctly aligned. This is done by placing a large pan at the bottom to catch the glassful of water you pour down from the top, watching as you do how the water runs over and around whatever you have in the bed. It is surprising how often the stream has been modelled so that it runs uphill. When the water has revealed the natural path of the stream, the next business is to model it. There are two main alternatives. If the stream is shallow, successive coats of clear varnish can be used. The second way is to mix very small quantities of two-part casting resin and pour them down the channel. A thin layer is formed each time, until the requisite depth is built up.

Waterfalls are slightly easier to reproduce. You take advantage of the facts that, when they are dry, both polystyrene cement and acetate cement are transparent and that both have a tendency to string. Taking the tube, apply it to the top of the fall and then, squeezing gently, slowly draw a long thread down your fall. Repeat this several times and allow the threads to dry. If necessary, add extra filaments to the fall. With what is left of the cement, create a sort of whirlpool at the bottom, and there you have it.

RUNNING WATER

While it is impossible to model running water, the effect of a mountain stream can be achieved by using casting resin, arranged to 'flow' over small 'weirs' made from model 'rocks'. Small stones from the garden can be embedded into the plaster.

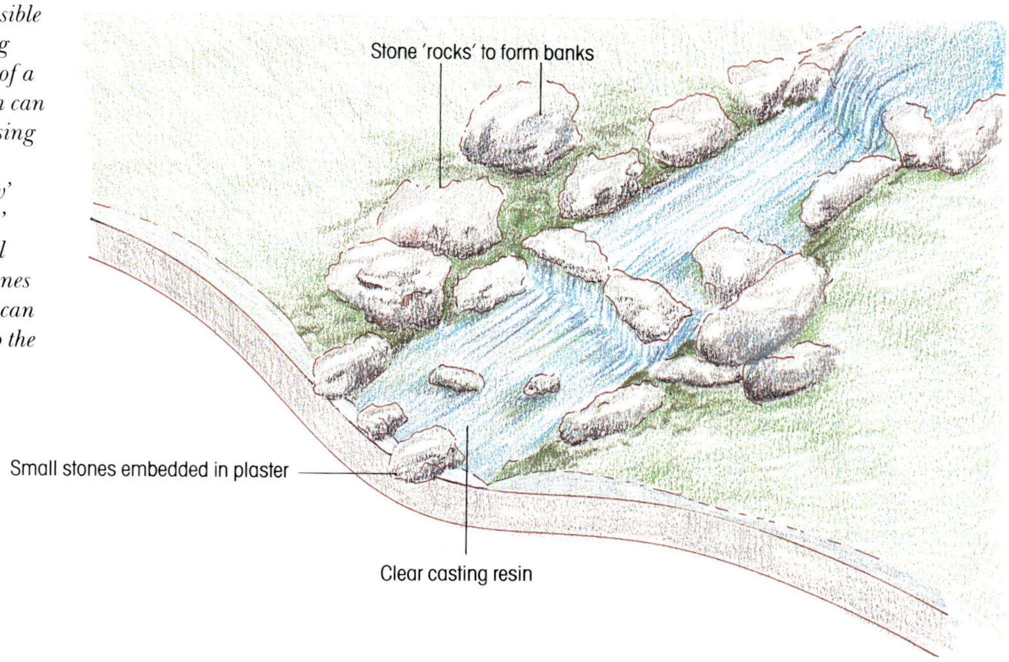

The sea is probably the most difficult of all to model, unless you opt for a flat calm. It is possible to create waves only by modelling the surface in plaster and then, when you are satisfied with the effect, to paint the surface and varnish it. It is probably best to prepare a few small trial surfaces to refine your techniques before beginning the main model.

When you are modelling landscapes it is a great help to have an illustration or two to hand as inspiration. Many guide books, albums, magazines and calendars contain superb colour illustrations, which should be kept for reference.

ROAD AND RIVER BASES

This diagram shows how the bases for roads and rivers can be arranged in an open framework. The framing is arranged at a lower level to allow for the modelling of a valley. The inset plan shows how the model would appear in its completed form.

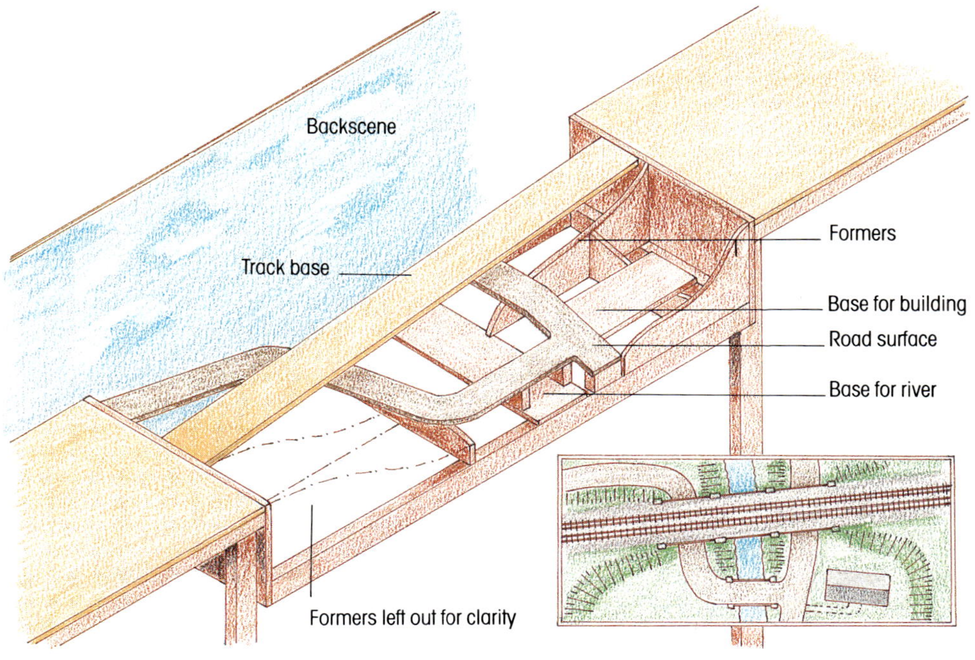

CHAPTER NINE

CIVIL ENGINEERING IN MINIATURE

With very few exceptions, railroads are not laid on the surface of the ground. Instead, a carefully graded road-bed is created on which the track is laid. Although it is convenient, in station areas at least, to lay model railroad tracks on to a level surface, in other places the ground should undulate to help create that all important illusion of reality.

For the most part, the tracks will either be on an embankment or in a cutting. On a model it is not unusual to find that most of the track is in cuttings, because it is quite easy to add hills on top of a flat surface. The snag with this is that trains are hidden, and it is better, on a model, to allow embankments to predominate. This is the main reason for adopting the open-top baseboard construction outlined in Chapter 5.

Embankments are invariably made by tipping spoil on to the ground, and they thus

BELOW

Manufacturers the world over offer a wide range of masonry and steelwork bridge kits for both HO/OO and N.

FACING PAGE

Timber viaducts are more popular on the model than they ever were in full size, where maintenance was a recurring problem and fire a potent hazard in the steam age. Whilst the trestle is more commonly associated with Western US practice, many wooden viaducts were used in Europe.

GWR

TARA E. QUMSY
Baby Goods
WILIKERS
ble Mint
Aid Teeth-Breath
Digestion-Millions Do
GUM
Coca-Cola
NIPS
5¢
L. M.

LEFT

A magnificent model of a Scherzer rolling lift bascule bridge dominates this section of George Sellios' HO gauge freelance Franklin & South Manchester Railroad. A 2-6-0 mogul crosses at the head of a rake of heavyweight clerestory-roofed passenger cars.

EMBANKMENTS AND CUTTINGS

Embankments and cuttings appear more realistic if two points are borne in mind – the slope of the banks, much shallower than is often realized, and the original lie of the landscape. Note the path for track gangs and the provision of ditches in cuttings. The railroad boundary is outside the earthworks.

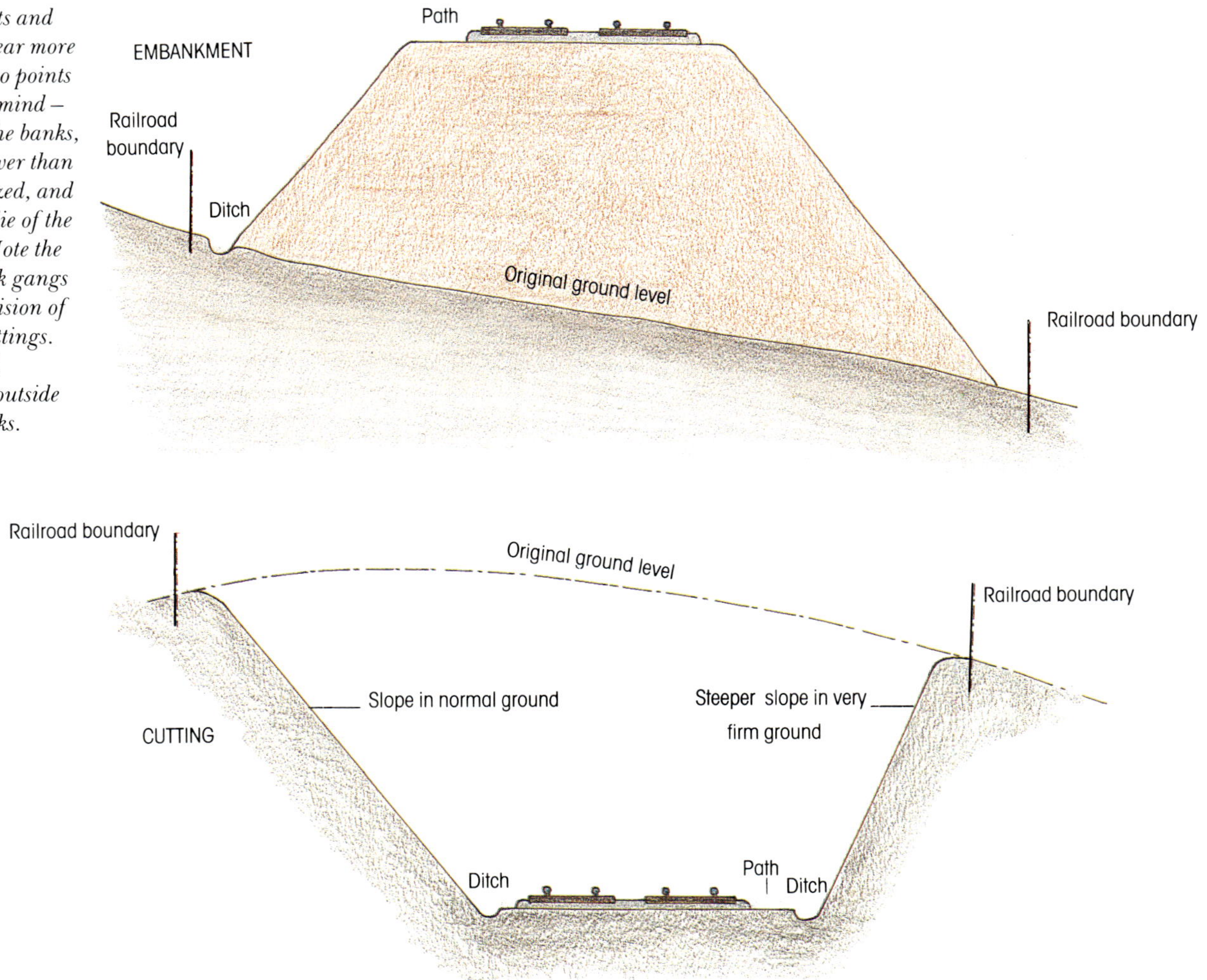

tend to spread considerably. Cuttings are excavated into firm ground, and so the slope depends very much on the type of soil encountered; chalk, for example, permits the cutting sides to be much nearer the vertical. As a general rule, for both cuttings and embankments you should aim to have the horizontal measurement roughly twice the vertical, although in most cases this has to be taken with a certain amount of modeller's licence. There are, of course, exceptions; in rocky terrain, for instance, the soil consists of rocks that not only have a steeper natural slope but are frequently built into freestone walls to support the track.

Railroad property extends at least to the base of the embankment or the rim of the cutting. Therefore, where the right of way is fenced this is carried along the boundary and not, as is occasionally done by modellers, close to the actual line of tracks.

Railroads cross roads, rivers and valleys, and so bridges have to be provided, while tunnels are needed to pass through larger ranges of hills and mountains. These form attractive features on a model, and a wide range of commercial products is available, mainly in kit form. Furthermore, being essentially utilitarian structures, they are not tied to any one country.

It would be easy to generalize and to suggest that while grade (level) crossings are the norm in the US, in Europe a bridge is preferred. The fact is that in much of the US, the railroad preceded the metalled road and because, at the time of construction, there was relatively light traffic on both routes, the cheaper arrangement was preferred. In Europe, by contrast, the roads were well established and usually crossed the railroad at a point where there was a cutting or embankment, so a bridge was the most

FACING PAGE, TOP
Anything with three or more arches is generally regarded as a viaduct rather than a bridge, but the distinction is a fine one and normally, as in this O gauge example, a viaduct carried a railroad line over a valley.

OVERBRIDGE AND RETAINING WALL

Where additional tracks are required in a cutting, it is often the practice to construct a retaining wall to gain the extra space without the need to purchase more land. After about 1870, new bridges across cuttings were frequently provided with side arches as shown.

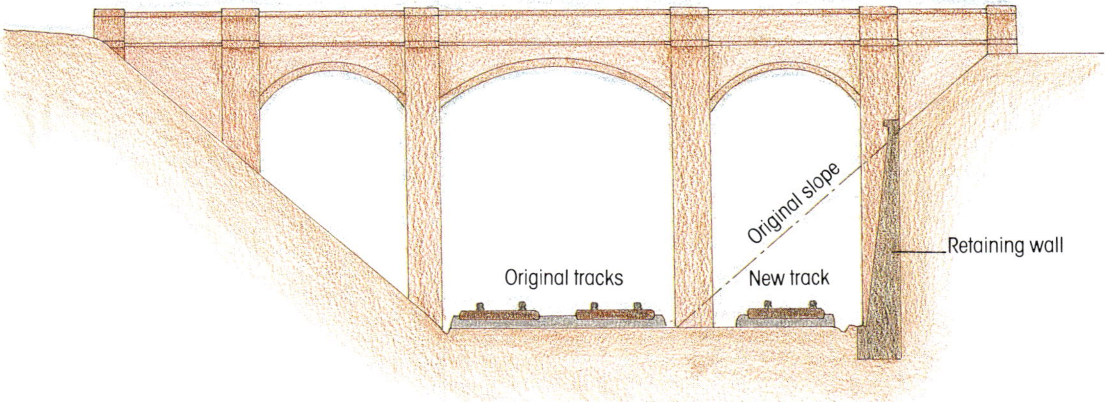

ABOVE

A straightforward masonry viaduct on an N gaugè layout.

RIGHT

A pair of tunnel portals neatly blended into the modelled landscape on a 4mm-scale gauge layout. A double-headed coal train trundles along on the main line whilst a Midland Johnson 0-4-4 well tank heads a four-coach train on the branch.

economical answer. Apart from this, bridges are provided wherever traffic is heavy and, today, where road speeds are high.

Bridges are also provided to cross rivers and deep valleys. Longer bridges are called viaducts, but the dividing line between the two is difficult to draw, although in general a viaduct is regarded as having three or more equal sized spans, while anything else is a bridge. Bridges can be built from wood, masonry or iron or steel. Few wood bridges remain on modern railroads, but their attraction is such that modellers love them. The best known type is the timber trestle, which is extensively used in the western states of the US, where the pioneer lines encountered large, well-wooded ravines and, because the basic material was close to hand, developed a simple structure that met its purpose admir-

BRIDGE CONTINUATION

Where one track passes over another near to right angles, a normal bridge or even a viaduct can be used (top left). Where the crossing is more acute, a skew bridge is needed (top right). Here, steel girders are preferred to masonry. However, on a model railroad such crossings often occur in the corner of the layout, and where both tracks are curved the problems multiply. Sometimes a series of girders, supported on portal frames can be used (bottom left), but a better arrangement is to put the lower track into a covered way – a shallow trough with a continuous deck above, carrying the high-level tracks (bottom right).

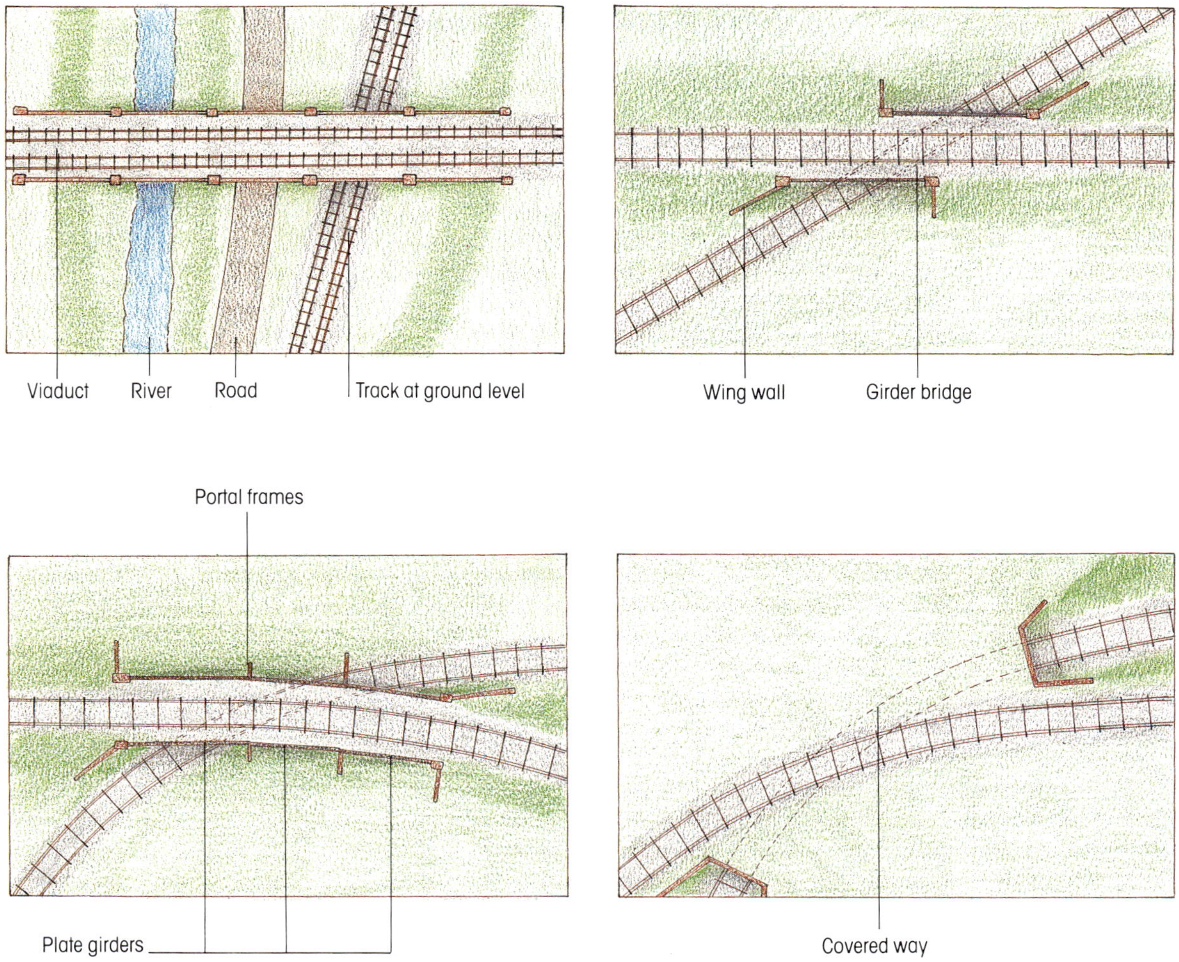

ably. In the early days in Europe timber was also extensively used, but here it was more common to use squared timber.

Kits for HO/OO and N gauge bridges are widely available, and patterns can be found to meet most requirements. Be wary of modifying bridge kits. Do not fall into the common trap of using two steel girder bridges end-to-end to form a longer span, for this not only produces a bridge so slender that it would collapse under its own weight were it enlarged to full size, but the bracing will be all wrong. As a rough guide, a plate girder has a proportion of at least 1:24.

A common problem on model railroads occurs when one track has to pass over another at a very acute angle so that the resulting bridge would be extremely long. This is often further compounded by the fact that

RIGHT

A Deutsche Bundesbahn BR 75 2-6-2 tank heads a mixed train across a masonry viaduct, which has been carved out of a block of expanded polystyrene. Note the beer van at the head of the train.

ABOVE

Road and rail traffic are concentrated around an underbridge on this excellent scenic feature.

LEFT

This structure – either bridge or viaduct, depending on your definition – has been scratchbuilt, using pre-coloured embossed card overlays to represent the masonry.

LEFT

A powerful BoBo metre gauge electric locomotive – one of two – is used by the Rhaetian Railroad to haul heavy trains through the long tunnels and over the steep gradients of the Swiss Alps. Bemo's excellent HOm (3.5mm scale on 12mm gauge track) model thus works well in conjunction with feats of civil engineering.

LEFT

A well-detailed coaster, complete with removable hatch covers and coal cargo, waits at the quayside. The classically English hump-backed bridge in the background is another delightful touch on this 0-16.5 narrow gauge layout.

GRADIENTS

The main object of the prototype civil engineer is to arrange as near a level route as possible, and to keep gradients as low as practicable, given the nature of the terrain and the resources available. If anything, the railroad modeller has a slightly different outlook, for lack of space, coupled with a widespread desire to carry one track over another for scenic effect, places more emphasis on the maximum gradient that trains can climb. Fortunately, most model locomotives will tackle far steeper gradients than their full-sized prototypes, but even gradients appreciably steeper than 2 per cent limit trains to four or five coaches.

Gradients on full-sized railroads are, in most of the world, expressed as a percentage. The exception is, or rather was, the UK, where they were until quite recently, shown as the length per unit rise, a 2 per cent grade being the equivalent of 1 in 50.

The real question facing the modeller, however, is not so much what the grade might be in theory, but how much room is needed to carry one track over another. This depends on two factors: first, the amount of clearance allowed between the two tracks, and second, the maximum length of train and the type of locomotives used. Taking 4 per cent (1 in 25) gradient as the measure, a 2m (6ft 6in) long grade is needed to give a reasonable clearance for HO or OO. This assumes one track is level. If both tracks are graded, the distance can be reduced to 1m (3ft 3in). However, this sort of grade will severely restrict train length, a total climb of 4m (13ft) for a single grade and 2m (6ft 6in) for a balance grade is to be preferred wherever possible.

in many cases the upper track is on a curve. When this happens on the prototype, the lower track is contained in a trough, which is then covered over by short cross-girders, the space between the girders being decked over to produce what is, in effect, a short tunnel, built on the cut-and-cover principle.

In model form tunnels are almost invariably fakes, in that there is no real bore. This is not a matter of laziness, but rather a recognition of the fact that it is necessary to provide access to the tracks for maintenance purposes. Where high-level sections are situated above lower tracks, make sure that the upper sections can be lifted clear; while it is, in any case, advisable to allow hand access to the stock so that any derailed vehicles can be easily removed.

Most tunnel portals are fairly functional structures; indeed, some bores through hard rocks are not even provided with a fascia. However, some quite elaborate façades were built, and as Faller has produced a kit for both of the elaborate portals of the Lorelei tunnel by the Rhine, quite a few layouts have acquired these extremely attractive adjuncts.

FACING PAGE
A pair of identical kits have been put together to form this double-track bowstring girder bridge.

RIGHT
There is little difficulty when a bridge crosses a track at right angles, as here. However, the problems multiply when both the bridge and the line below are curved – typically, this will occur in the corner of a layout.

ARRANGEMENT OF GRADIENTS

If only one track is on a grade, a fair distance is required to allow one track to pass over another – in HO/OO gauge at least 6 ft 6 in (2 m). However, by having one track rise and the other fall, the distance can be halved.

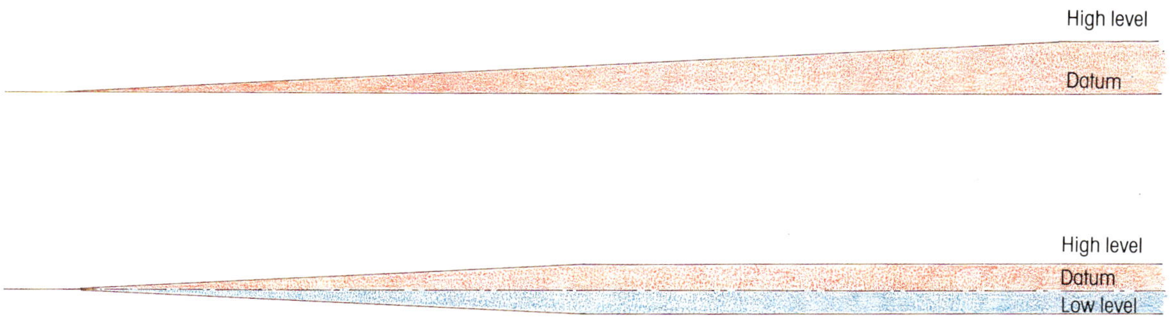

CHAPTER TEN

MODELLED ARCHITECTURE

Buildings fall into two clear categories – the railroad structures proper and those buildings found beyond the railroad's boundary. Since the former are, in the main, functional structures, they tend to be fairly similar throughout the world, and although there are national characteristics, these are not infrequently matters of detail rather than fundamental differences. Indeed, there have been instances of deliberate copying; for example, the wooden intermediate station buildings on the West Highland line in Scotland were officially described as 'Swiss chalet style'. Whether this was done as a ploy to attract tourists or in the belief that, in an area where snow was fairly common, a Swiss design would prove effective is difficult to determine for sure.

When one moves outside the railroad's boundaries, national styles of construction are apparently of paramount importance. Certainly, there are pointers – roof styles, for example – and since, on a model, these features are very obvious, this can be very significant. However, factories and warehouses are much the same all over the world, the style being more dependent on the date of construction than on the country of origin, and modern architecture has few national characteristics. When we turn to major pub-

BELOW

A range of station building kits by Faller. The model in the foreground is based on the main station at Bonn, Germany.

FACING PAGE

An impressive model building, built from a Vollmer plastic kit. Whilst based on a German original, this type of structure is to be found all over the world and is more typical of its period (1890-1910) than its country of origin. Though shown as an hotel, it could equally be a block of apartments, offices, a museum, an art gallery or even a town hall.

Grand Hotel

ABOVE

A lovely example of the effective use of a number of small building kits to create a convincing representation of a medieval walled town. Although the area modelled is far too small, the grouping of ancient buildings around an irregular square is pleasing to the eye.

lic buildings – the railroad station itself, town halls, hotels and so on – there is a considerable affinity between designs from the 19th century onwards. Churches are perhaps more traditional in design, but once again, the 19th-century Gothic revival style crosses national boundaries, as indeed did the original Gothic of the Middle Ages.

Most model buildings are sold in kit form for various reasons, not the least important of which is that, when reduced to flat walls, even a massive structure can be put into a relatively thin box and so ease the stockist's space problems. There is also the point that it is a lot easier to manufacture relatively thin, flat components. Today, most kits are moulded in plastic, usually in several colours, so that there is no need to paint the model afterwards. Some years back a hybrid type of kit was introduced. These had self-ahesive printed overlays, but they were not wholly popular for not only did the model lack surface texture, but putting the sticky sides accurately in place, without wrinkles, proved rather more difficult for the user than the designers believed. The plastic building kit is very much a European speciality, and the products of the leading firms are distributed worldwide.

A few kits are made from printed card, and some of these are pre-cut. In the main, however, this type of kit is produced by smaller firms and is aimed at the British market, rather than at modellers in other countries. A very few wooden kits are produced. These are intended for craftsmen, and, in general, the ranges are fairly short-lived since anyone who is capable of assembling one of these structures is also capable of building the model from scratch.

Because buildings come in so many shape and sizes and are then, over the course of years, modified by successive generations, this is a field where building from scratch pays dividends. However, before we look at this aspect, we should consider kits because they offer the best entry point into the craft of miniature architecture.

Card cut-outs are frequently sold for children because a simple, brightly coloured card kit can be produced very cheaply when printed in quantity. Indeed, some 20 years ago, many breakfast cereal boxes had card cut-outs printed on the reverse. (Incidentally, cereal boxes are an excellent source of suitable card for modelmaking.) The sad truth is, however, that although it appears to be simple enough to cut out a sheet of card, fold it along the dotted lines and glue the tabs together, making a good model in this fashion is far from easy.

Assembling a plastic kit on the other hand, is reasonably straightforward. The most important point is that the instruction sheet should be studied carefully before assembly is begun and that parts should not be taken from the sprues (the waste sticks formed in the mould when the plastic is run in) until you are ready to put them in place. This is particularly true of the smaller parts, since the all-important identification number is often embossed on the adjacent sprue. Assembly is best done by applying a plastic solvent with a brush rather than by using plastic cement direct from a tube. When a solvent is used, the parts are put together dry, and, in most cases, a small amount of the fluid is applied to the joint with a fine brush. The fluid then flows into the gap by capillary action. The joint is strong enough to hold firm in about 5 seconds, although it takes over an hour to dry completely and so a mistake can be rectified within 10 minutes with little difficulty and within the hour at the worst. Since errors usually become apparent within a minute, this makes life much easier. In most cases, too, the solvent can be applied on the inside of the joint.

Window and door frames should be assembled first, and glazing and 'curtains' applied in the flat. The walls are then put together on the base, and a generous supply of solvent applied to the inside. It can be helpful to apply a small fillet of plastic cement along the main joints when assembly is complete. Roofs, which are usually quite vulner-

ABOVE

Effective use of foreground buildings and printed backscene creates the illusion that one is looking at a large town, stretching far out into the distance.

ABOVE

The impressive depot on Rudi von Prittwitz's HO gauge Southern Pacific layout. A silver Budd RDC (Rail Diesel Car) sits in the foreground whilst in the rear a Santa Fe 4-8-4 heads a rake of clerestory-roofed heavyweight passenger cars. Whilst most of the buildings are from kits, the impressive main station structure is a scratchbuilt model of Merced station, California.

able, often need a small fillet of cement to hold them securely in place. External details are applied last of all. It is a good idea to spread the assembly of the larger, more intricate kits over a couple of evenings so that the joints can harden off fully before the next stage is begun.

Most plastic kits are self-coloured, and, in theory at least, they do not require painting. This is, to some extent, true of the better quality models, since the manufacturers take considerable trouble to provide a reasonably matt finish to the parts. Furthermore, painting a kit-built building often leads to a very unsatisfactory finish, particularly if it has been carried out as something of an afterthought. Some of the cheaper kits, which are moulded in one colour plastic, must be painted.

LOW-RELIEF MODELS

We see only one, or at the most, two sides of any building at any time and so, for use along the backscene and in the further parts of the model, low-relief modelling enjoys a considerable vogue. These structures can be as little as 3mm (⅛in) thick and consist of no more than the facing wall of the building. In such cases, you must either construct a 'flat' roof or model the gable ends. Most low-relief buildings are at least 1cm (½in) deep, allowing a fair representation of the sloping roof to be simulated, even though the angle is far more acute than one would normally see in practice. In the extreme cases, a standard building is, in effect, cut in half. This is a relatively simple exercise with most kits, and all that must be done in many cases is to cut the end walls cleanly in half, for the roof is already in two pieces, and the rest of the kit remains as is.

The deeper models allow the inclusion of shop windows and other detail in depth, and also provide space for lighting.

One of the great virtues of low-relief modelling is that it produces more effect for less work, and saves space, time and money.

BELOW

This view of a row of low relief structures show how only the façades are fully detailed. This type of model is often situated at the back of a station to create the impression that the railroad serves a large township. Detailed shop fronts are a popular theme for this type of model.

It is a lot easier to paint the parts before they are assembled and while they are still attached to the sprues, which are a convenient way of holding the parts while the paint is still wet. The main difficulty remains that of applying a flat coat of paint, since the faintest hint of gloss will spoil the effect. Oil-based paints are particularly prone to this. One solution is to drain off the majority of the oily medium from the tin, leaving a cake of moist pigment behind. Put a small quantity of this on to a palette and apply it with a brush moistened with pure turpentine or, if this is not available, with white spirit. The palette can be any suitable surface such as an old saucer. The turpentine can be held in a small, open container – the metal screw top from a soft-drink bottle is perfect.

ABOVE

Kit manufacturers do not confine themselves solely to buildings, as this view of the Faller 'Nodding Donkey' oil wellhead pump illustrates.

DANGEROUS LIQUIDS

Plastic solvents, turpentine and white spirit fall into the large group of flammable liquids used in modelmaking. They can be used carelessly for years without undue incident, but it takes only one mischance to start a fire. Two simple precautions should be followed at all times.

The first is to store all such materials in metal boxes, so that, should fire break out, the fluid is confined. The real danger occurs, however, when a burning fluid spreads over the floor and sets fire to everything in its path. If they are kept in a metal box, it is often possible to pick up the box and throw it out of doors, or even to smother the fire by placing a rug or thick blanket over it. It is also possible to direct the contents of a fire extinguisher into the box.

The second precaution is to decant the fluid into small bottles for immediate use, and to store the bulk supply safely in an outbuilding rather than in the house. It is not necessary to fill the bottles. If you have enough for a couple of modelling sessions only, there is that much less to spill. One difficulty with most small bottles is that they are inherently unstable and readily knocked over. There are some suitable designs around – low, squat, sometimes triangular-section bottles – and it is worth keeping a look out for suitable patterns. Failing this, either make a wooden block holder or keep the bottles on a small metal tray, since, quite apart from the fire risk, a spillage over the workbench means that all parts get soaked, and this does not improve any plastic kit.

It is also essential to point out that all liquid solvents are potentially toxic. In particular, the fumes are dangerous and can easily cause severe headaches. It is essential to work in a well-ventilated space, and it is advisable to wash one's hands thoroughly after a modelling session, since dermatitis is easily contracted from prolonged contact with common cleaning fluids, let along the more exotic substances. The risk is small, but should be kept in mind.

You will also need a quantity of white spirit for cleaning purposes. A small bottle, with about 1cm (½in) of spirit in it, is needed to clean the brushes immediately after use, not at the end of the session. Give them a quick swirl in the fluid, followed by a thorough wiping with a piece of rag. A further clean at the end of the session is advisable, and at the same time surplus paint from the palette should be scraped away and discarded and the palette wiped clean with a rag dampened in white spirit so that it is clean for the next session.

The dry brush technique, by which the pigment is taken up with a barely damp brush and dragged over the surface, is particularly useful. When it is applied to a rough plastic surface, such as moulded brick or stone, it quickly produces the desired uneven finish. For this, a narrow, 4mm (⅙in) wide, square brush of the type supplied for artist's oil painting, is most suitable. For general purposes the conventional small brush, sizes 1 to 3, is most suitable. Brushes should be top-quality modelmaker's brushes or middle to top quality artist's brushes. Cheap brushes are worse than useless. The very fine (O or OO) brushes are useful for applying touches of a second or third colour to selected bricks or stones, a process that is a good deal easier in practice than it sounds on paper. Top-quality brushes must be used because cheaper varieties do not form the fine point needed for this delicate work.

Although it is easier to paint the parts before assembly, some details – quoins (corner stones), for example – need to be painted when the model is assembled. It is often necessary to add a few other final touches at this stage. Lettering for shops, offices, factories and so on is supplied with most kits, although the range is, of necessity, limited. Separate plastic letters in varying sizes are obtainable from the better modelshops, and these make it possible to provide a good range of raised signs with relative ease. In addition, rub-down lettering is now widely available, and some specialized suppliers provide ranges of lettering as decals (waterslide transfers). These last items can be rather difficult to track down and when found, should be stocked up in bulk since they are no trouble to store.

Weathering the outside of buildings must be carried out after assembly. The

BELOW

Heavy industries, which rely on large quantities of raw materials, are often found clustered by railroads on the outskirts of towns and cities. This oil storage depot gives a gritty, urban feel to the layout and also provides a justification for special wagons!

simplest method is to apply a thin wash of dirty brown paint sparingly to the exposed surfaces. You will find a very suitable solution in the bottle you used to clean your brushes! Look closely at full-sized buildings and note how rain streaks cause vertical stains; obviously, do not do this if your model is set in an arid part of the world.

So far we have assumed that the kit will be assembled according to instructions, and this should, indeed, be the rule at the outset. However, plastic kits lend themselves to modification, and a good deal of variety can be produced by using parts from two or more kits to make an original model. It is also possible to combine several kits into one single unit, and this is necessary if one wishes to model the older, central parts of a town, where it is usual for different buildings to be built into each other. It is possible to take two assembled buildings and to link them with parts of a third, following the infill process so common in most towns.

When plastic kits are used in this fashion, not only is it necessary to cut walls, roofs and other parts to suit the new model, but it is also useful, from time to time, to use plastic sheet, as sold by most good modelshops, to supplement the kit parts. The better modelshops can also supply moulded and embossed plastic sheet to simulate all common, and not a few uncommon, building materials. Specialist suppliers provide, either as plastic mouldings or metal castings, a range of architectural details ranging from windows and doors to chimney pots and roof finials. Some parts are produced in etched metal, and railings, in particular, are often made in this fashion.

The point at which a kit has been so extensively modified that it becomes a new model in its own right is difficult to define. It is not, in any case, of interest outside the specialized area of the competition modelling, where considerable weight is given to the supposed virtues of scratchbuilding.

However, when dealing with such a varied subject as a building, scratchbuilding

ABOVE

As always, attention to detail brings colour and interest to an otherwise humble rural scene. Note, in this instance, the bicycles, rubbish cans and mail bags on this OO gauge layout of Brackenbridge Station, UK, set in the 1960s.

has one obvious merit: the modeller is free to copy any structure that takes his or her fancy. There is the added virtue that the modeller working in this way has a wide choice of materials available. Cereal packets have already been mentioned, and these and other items of packaging have been successfully used for modelled architecture for many years. The relatively thin cereal packet requires a good deal of reinforcing, and thicker card may be found in the outer cartons in which the pre-packaged goods arrive. Since cardboard is formed of several layers of material, glued together under pressure, it is possible to produce a good deal of relief by partly-cutting through the material and peeling successive layers away.

Wood, generally in the form of thin ply, is a very popular material that forms a good foundation for larger buildings. It is also an excellent medium for modelling all-concrete buildings. Modern all-glass buildings are modelled from acetate sheets – Perspex, for example – with superficial details added. However, Perspex is a tricky material to use as the only readily available solvent is chloroform, and this is not an easy product to buy in most countries. It is not exactly pleasant to use, either!

Wood or card buildings may be painted, but in most cases 'brick' paper (assuming, of course, that this conforms with the style of architecture) is preferred as a final finish. This material is less common than it used to be, but it can be obtained from the better specialist modelshops and some art shops. It is best applied with a rubber-based mounting adhesive rather than paste or glue.

ABOVE

A very impressive model of a main station building, in the monumental classical style favoured by British architects in the early decades of railroad development, graces the main station on the Halifax Model Railway Club's OO gauge layout.

ABOVE

This layout of York Haven Rr. in HO gauge shows how best to arrange a townscape behind a model station. First, the buildings are raised above track level, ensuring that the view is not obscured by the trains. Secondly, the retaining wall makes an excellent foil for the trains. And finally, the space under the town can be profitably used for train storage roads, which are readily accessible by lifting the street scene clear off its supports.

LEFT

Plastic kits form this small townscape, neatly filling a corner of Bill Roberts' N gauge Geislingen layout.

CHAPTER ELEVEN

NARROW GAUGE

Narrow gauge railroad modelling began as a minority interest centred on systems that were all built from scratch. It has developed to the point where there is now a considerable amount of ready-to-run equipment available, not merely in specialist modelshops but frequently in the larger high street toyshops as well. Because there are, on the prototype, so many gauges to choose from, narrow gauge nomenclature becomes a trifle complicated and, for precision, the best system of identification is to state the gauge in millimetres and the scale adopted. Commercially, there is a distinct difference between US and European practice.

In the US, the practice has been to select the scale and then lay tracks to the true gauge, hence the popular HOn3 system, 3.5mm scale (HO) on 10.5mm gauge track (true 3ft). This has been developed commercially by the introduction of accurate track and turnouts and an array of top-quality, ready-to-run brass stock, mostly of Far Eastern manufacture. In Europe, the practice has been to adopt an existing gauge and to choose the most appropriate scale, a process aided by the fact that, by a happy chance, 12mm gauge (TT) is exactly 3ft in 4mm scale, and not far off 1m in 3.5mm scale (1:87, HO), while 9mm gauge, with either 3.5 or 4mm scale suits the various odd gauges around 2ft.

Commercial development began with HOe, 1:87 scale, 9mm gauge, with tiny four-wheeled steam- and diesel-outline loco-

RIGHT

The Rhaetian railroad, an extensive metre gauge system in the south-east of Switzerland, has grown in popularity amongst narrow gauge modellers in recent years. This is due, in part, to the spectacular nature of the prototype but mostly to the enterprise of Bemo, who now offer all current RhB locomotives in ready-to-run form, together with a very wide selection of coaches and wagons. Here we see their model of the 1920s-built 'Baby Krok' articulated locomotive at the head of a goods train.

FACING PAGE

The spectacular nature of a 3ft gauge railroad is highlighted by this On3 ($\frac{1}{4}$in scale, 3ft ($\frac{3}{4}$in) gauge) Crown Mountain Railroad. The locomotive is a Colorado Southern 2-8-2, and a Baldwin saddle tank switcher shunts the mineral siding above.

ABOVE

Grasmere station, constructed on a 0-16.5 (7mm scale on 16.5mm gauge track) narrow gauge layout. The model is set in the English Lake District, with stock on a variety of British prototype narrow gauge lines. The locomotive at the head of the passenger train is an ex-Glyn Valley 0-4-2 tank, whilst the slate train on the left is characteristic of the region.

motives and a selection of 'typical' coaches and wagons. In the UK 4mm scale was used with the same stock, since the many manufacturers of narrow gauge equipment offered their standard equipment in a range of sizes. There was some support for 3ft gauge modelling in Britain; this was entirely based on kits and is outside our immediate terms of reference. In recent years HOm metre gauge has flourished.

The other narrow gauge development of note is G gauge, 45mm track and, initially, a 1:22 scale for 60cm gauge, but later developments include 1:32 scale models for metre (Swiss) or 3ft (US) gauge. The initial products by LGB are fairly costly, but they are rugged and excellent performers. Recently, the Playmobil range, which is offered as toys, has provided a cheaper entry. G gauge is at

PROTOTYPE NARROW GAUGE

Narrow gauge railroads are technically those with a gauge less than 1435mm (4ft 8½in), but as the normal gauge in many parts of the world is either 3ft 6in or 1m, narrow gauge is generally used to refer to feeder railroads with a smaller gauge than that current in the rest of the region.

In Europe and the US, prototype narrow gauge falls into two distinct categories. First, there are the secondary railroads, which are 3ft gauge in the US and metre gauge in Europe. Although some lines built to these gauges were short feeders, the principal use in the US was a secondary system for use in mountainous and sparsely populated areas, while in Switzerland there are several large networks over which trains run behind powerful modern electric locomotives. Similarly, in the US, large steam locomotives and long trains of commodious, period coaches are the order of the day.

Then there are the smaller narrow gauge lines with gauges from 60cm (2ft) upwards. These are mostly feeders, and many, like the pioneer Festiniog Railway, originated as mineral lines, often in connection with a small port. The survivors are mostly tourist railroads.

RIGHT

Very large narrow gauge. LGB G gauge locomotives, based on standard products by German steam locomotive builders. These massive models are normally used in garden settings, but the tight curves do permit simple lines to be built indoors.

its best in a garden setting, where the large-scale models are not seriously out of proportion to the plants.

The main advantages of narrow gauge, both full sized and model, lie in the fact that much sharper curves are possible. On the prototype there are also savings in construction costs, although this is not always the case on the model. However, in model form, narrow gauge does lend itself to 'fun' layouts, where modeller's licence can run riot. In addition, on a more serious note, narrow gauge offers the chance to get an interesting system into a relatively small space. It is, above all, an option with considerable scenic appeal, since many narrow gauge lines have been built in mountainous terrain, where there is a need for spectacular bridges, frequent tunnels and, above all loops and spirals carrying the tracks up the steep valleys and providing the modeller with a valid excuse for all those exciting arrangements that produce a visually interesting layout.

Principal Narrow Gauge Sizes

Gauge	mm	Scale	Scale/foot
G	45	1:22	—
1m	32	1:30.5	10
SM32	32	1:20	16
016.5	16.5	1:45	7
OOn3	12	1:72	4
HOm	12	1:87	3.5
HOn3	10.5	1:87	3.5
OO9	9	1:72	4
HOe	9	1:87	3.5

BELOW

Bugsworth on the Hazel Grove club's narrow gauge High Peak Tramway. Minerals, probably limestone, have just been tipped into the waiting canal narrow boat.

BELOW

This scene on the Hazel Grove High Peak Tramway is typical of much narrow gauge modelling: a busy scene around the simplified locomotive maintenance facilities, represented by the small shed and its environs. A large water tank spans the front of the shed, an oval tank, taken from a small road tanker, supplies fuel oil to the locomotives.

CHAPTER TWELVE

THE NEXT STEPS

The methods described in this book will enable you to construct a very satisfying model railroad. Human nature being what it is, however, most of us will want to improve on our early efforts in this, as in any other field. One thing you should take out at an early stage is a subscription to the leading model railroad magazine in your own country, for, apart from anything else, this will tell you where to get hold of the more elusive items that are not available at your local stockist. It will also show you the work of other enthusiasts, and these articles and photographs will inspire you to further efforts.

You need to be a little cautious about trying to emulate the layouts shown in the magazines. It is said that the camera cannot lie, but in competent hands a single lens reflex can create the illusion of reality as long as the viewpoint is low enough. Careful photography can also help to hide the fact that the model might be very small and should be regarded more as a diorama than a complete model railroad. There is not, of course, anything amiss with diorama modelling. If you take an area some 2 × 0.8m (6ft 6in × 2ft 6in) and spend a year making a fully detailed model of a section of landscape that includes a railroad line, the end result should be a visually attractive model. Dioramas are good fun, and an excellent way of dealing with severe lack of space although at the expense

RIGHT

Approaching the ultimate in modelled realism, this O gauge (7mm scale) GWR Dean Goods at the head of a freight train is a superb example of what can be achieved by meticulous individual craftsmanship.

FACING PAGE

The charm of large scale narrow gauge modelling is brought out in this superbly atmospheric shot taken on this Gn3 (45mm gauge LGB) Soda Creek & South Park layout. The weathered locomotive, spindly trestle and finely worked trees are testaments to the modelmaker's art. The final touch of realism is provided by the mist effect in the valley.

LEFT
Signals add to the realism of any model railroad, as this shot of an OO gauge layout clearly demonstrates.

of operating potential. They are not the whole of the story, though.

Another popular approach is the construction of locomotives, coaches and wagons, either from kits or from raw materials and a few difficult-to-make components. You can spend a whole year building one superbly detailed locomotive, while a complete rake of coaches, each with fully detailed underframes and full internal fittings – including passengers – can take even longer. You could even set to and a model an actual building down to the last cracked brick, again with full internal detail if you feel so inclined. If you have the right temperament and are

CLUBS AND SOCIETIES

It has often been said that the aspiring railroad modeller should lose no time in joining a model railroad club. Certainly, most established club members praise the system to the skies, even if they may be highly critical of the shortcomings of their own society. However, it is probably the case that they enjoy club life, and it is easy to forget that not everyone does. To be a good club member you must be prepared to put time and effort into the organization, and at the very least, you have to turn up regularly and take part in whatever is going on.

There are several types of model railroad club. Some exist primarily to build and operate a joint layout. Such groups are invariably set to one specific scale/gauge combination. They are, in addition, careful about accepting new members, for unless it is clear that the individual concerned will not only conform to the established norm but will also contribute positively to the group's work, he or she will be an encumbrance. Such groups are usually fairly small in number and relatively costly to join.

Another type of club embraces all scales, gauges and interests and accepts virtually anyone who cares to apply for membership. Although the organization's constitution may contain high-sounding aims, in practice such a society exists to hold meetings at which members can get together, exchange information, show off their latest models and attend talks, slide shows and so on. In addition, the club usually organizes an annual exhibition, which not only helps to supplement the club funds but also lends a strong sense of purpose to the proceedings.

Another type of association is the specialist group that caters for a specific scale/gauge combination and these groups are usually organized on a national basis. They publish newsletters or magazines, provide information and product support and organize regular meetings and exhibitions. Passive membership of such groups is profitable, as the journal and product support provided are the main benefits of membership.

LEFT
Detailed scenic modelling takes time and, when carried out as effectively as this, brings a model into the advanced category. The creation of a scene like this calls for both close observation of the original and a good deal of controlled imagination – it is without doubt an art form.

prepared to acquire the necessary dexterity in the use of tools, fine craftsmanship is extremely satisfying, and at the same time you are liberated from the limitations of readily available commercial models.

An electrical expert has been described as a man who uses a computer to do the work of an on-off switch. A more accurate description is someone who can understand and prepare circuit diagrams and who finds the business of wiring both absorbing and soothing. Given those preconditions, a model railroad offers endless fun, for there appears to be no limit to the complexity of circuits that can be devised, and there is now the choice between relays, in all their various forms, and solid state circuits, in their even greater variety. Furthermore, add to these the various indicators and meters that can be attached to the model, and it is easy to see that the potential is enormous.

Signalling – a topic we have not even mentioned – is also a fascinating subject in its own right. A signal expert has been defined as someone who does not care if table tennis balls run along the track, as long as they obey the signals. This is, of course, an extreme view. Signals are, nevertheless, fascinating and are essential for the overall realism of any layout. They can, however, be rather frustrating because not only does the practice of each country differ in overall principle, but individual companies often set their own standards.

These are only some of the more common themes that modellers pursue; there are sub-divisions within each grouping and some rather esoteric sidelines as well. Furthermore, you can indulge in any aspect of the hobby to any degree you like. As a result, everyone ends up with a purely personal layout. In most cases, too, we take it in turns to concentrate on the aspect of the hobby that happens to attract us at a particular time.

It is, however, all too easy to forget that the reason we went to all this trouble in the first place was to be able to run trains over our own layout. The operation of a soundly built model railroad is a never-ending source of delight, and it is fair to say that the majority of railroad modellers who have got their layouts into good order spend more time running the model than anything else. Indeed, no matter what various sectional interests may claim, the ultimate heights are scaled when, and only when, it is possible to go into the railroad room, switch on the power and then recreate, for your own pleasure, the operation of a stretch of railroad line. Ideally, this will transport you, and any kindred spirit with you, safely into a golden world that does not quite exist.

FACING PAGE
Exhibitions form an important part of the club scene and are an excellent way for the newcomer to see for himself just what is on offer. The oldest show of all, IMREX, held annually at Easter in Central London, provides an important showpiece of the hobby.

CHAPTER TWELVE

SCALES, GAUGES AND STANDARDS

There are a large number of scale/gauge combinations in existence, many of which are of academic interest as they are used only by a handful of specialists. The principal sizes are listed below and the notes below give some indication of their potential.

PRINCIPAL MODEL RAILROAD SIZES
STANDARD GAUGE

Gauge	*mm*	*Scale*	*Scale/foot*
1	45	1:30.5	10
O	32	1:43	7
OO	16.5	1:72	4
HO	16.5	1:87	3.5
N	9	1:160	—
Z	6.5	1:200	—

The largest standard gauge in use for model railways, gauge 1 used to be regarded as obsolescent, being largely maintained by the efforts of the British Gauge 1 Association. However, nearly 20 years ago Märklin revived the scale, and since then it has enjoyed a renaissance. Indeed, it is safe to say that more enthusiasts are either collecting or modelling in gauge 1 today than used the scale in its pre-1914 heyday. It is mainly used for out-of-doors layouts, for above all, gauge 1 is the scale for live-steam traction. It is the model engineering scale *par excellence*, and although the units are costly, both in time and money, a very little gauge 1 goes a long way.

O GAUGE

One commentator said that, like the waltz, O gauge is always on the point of coming back. Although commercial support, in the form of ready-to-run equipment, is very patchy, O gauge is fully supported with kits and components. The relatively large size makes it ideally suited for scratchbuilding, and there is the pleasing feeling that it is highly unlikely that, after one has spent a year or more making a fully detailed, true-to-scale model of a locomotive, someone will put an even more detailed mass-produced model on the market.

OO GAUGE

This size is a purely UK phenomenon; its origins lie in the fact that British locomotives, although running on the same track gauge, are smaller in cross-section than those in Continental Europe or the US, so that a 4mm scale model of a British locomotive is roughly the same size as a 3.5mm scale model of a Continental locomotive. Widely supported with ready-to-run stock and flexible track, together with a wealth of kits and components, it is the major system in Britain. The fact that HO and OO stock run on the same track is quite an advantage.

HO GAUGE

This is the most popular size in the world, with the further advantage, from a purely theoretical standpoint, that the scale/gauge ratio is as near exact as is possible. Except for British-outline models (see OO gauge above), it must be the first choice.

N GAUGE

Although experimental work was carried out in sub-HO gauges in the early 1930s, it was not until some 30 years ago that N gauge became established commercially. It offers considerable advantages if space is limited, but the gauge suffered initially because it was used more as a last resort than as a first choice. Now that a wider range of ready-to-run models is available, it is growing in popularity, not merely for small, portable systems, but for large layouts where one can run prototypical trains and produce complex layouts with several stations and alternative routes within an affordable railroad room.

Although some superb model locomotives and coaches have been scratchbuilt in this scale, it is not really suited for this aspect of the hobby. On the other hand, it is ideal for scenic modellers and for anyone who is interested in large architectural effects. Indeed, the standard architectural modelling scale of 1:150 is close enough to allow models originally intended for this purpose to be incorporated into an N gauge layout.

In the UK a slightly larger scale is employed, 1:148 rather than 1:160.

Z GAUGE

This is the smallest commercial gauge and is in the main only obtainable from one manufacturer, Märklin. Although it is attractive for sub-miniature projects, it does allow some spectacular effects in a moderate sized space. The twin disadvantages are cost and a very limited choice of complete models, which are almost entirely confined to Continental European prototypes.

The gauges described above are those most commonly used today, and are those that have strong commercial support. Numerous other combinations have been used, and although some combinations are obsolete, in the sense that they exist mainly as collector's items (pre-war Märklin OO gauge models are a case in point), others exist as specialist systems and are generally supported by enthusiast groups. Since these scales are almost entirely confined to one country, they fall outside the scope of this book.

GLOSSARY

Backscene A vertical board placed behind a section of the layout, with a painted or printed scene to suggest that the model extends into the distance. In its simplest form the backscene is painted light blue to suggest the sky.

Ballast On the full-sized railroad, the crushed rock placed between the sleepers (ties) to provide drainage and hold the track in place. In model form it is purely cosmetic.

Baseboard The framework on which the model railroad is built.

Bogie A four- or six-wheeled frame, a pair of which are pivoted underneath a locomotive, coach or long wagon to allow the vehicle to negotiate curves.

Boxcar (US) A covered bogie wagon.

Bullhead track An obsolescent type of track in the UK and France. The rails have an hour-glass section and are held upright in chairs.

Catenary The complete assembly of wires for overhead current collection; the term derives from the curve formed by the upper supporting wire.

Cattle wagon A van adapted for the carriage of livestock, provided with open slats for ventilation.

CCT (UK) Covered Carriage Truck – a van with end doors for loading cars, vans etc. Also used for general merchandise when required.

Chair A cast-iron fitting to secure the bull-head rail to the sleeper (tie).

Clerestory A raised central section of a carriage roof, provided with side lights to give additional light and air to the interior of the coach. Used on top-quality stock from around 1875-1900, after which a higher roof was preferred. The term is taken from architecture and is pronounced clear-story.

Crocodile wagon (UK) A low-slung bogie wagon with a depressed central well for the carriage of large indivisible loads.

Crossing (1) The intersection of two tracks without any interconnection, often called a diamond crossing from the shape of the rails. (2) The place where a road crosses the tracks on the level.

Crossover A formation of two turnouts linking a pair of parallel tracks.

Diesel The generic term for a locomotive powered, in full size, by a compression-ignition engine. Diesels fall into three groups: diesel-electric, the most common, diesel hydraulic and diesel mechanical. The latter type is confined to low-powered shunting units and railcars.

Diorama A scenic model, generally relatively small and, in its purest sense, designed to be viewed from one position. It is nowadays often applied to models of railroad scenes which are lovingly built to be looked at and photographed rather than operated.

DMU Diesel multiple unit. A set of two or more diesel-powered railcars.

EMU Electrical multiple unit, as DMU but powered by electricity.

Fiddle yard A series of tracks placed offstage, where trains are held, re-formed and brought out in accordance with the operating schedule. As this is not part of the imagined model world, no scenic features are added and tracks are not ballasted. In addition, locomotives and rolling stock can be lifted off the tracks and repositioned.

Flyover Where a track or tracks are carried over the main route by means of a bridge to avoid conflicting movements over a junction.

Gauge The distance between the top inner face of the rails.

Gondola (US) An open bogie wagon.

Gradient A sloping section of track. In a model it is normally employed to carry one track over another.

Hardboard A man-made board, ⅛in (3mm) thick, usually smooth on one side and patterned on the other.

LEFT

Not surprisingly the Vancouver Island-based North Island Model Railroaders have taken CP Rail as their inspiration for their HO gauge club layout. The locomotive is a standard SD40-2 hood diesel unit in CP Rail colours. Note the very realistic rockface in the background.

Insulation board A semi-hard man-made board, approximately $\frac{3}{8}$in (9mm) thick, often known (by its brand names) as Sundela in the UK and Homosote in the US.

Interurban An electrified light railroad linking two or more communities in the US.

Layout The overall term for a complete model railroad. Technically, it does not include locomotives and rolling stock, though these are essential if the layout is to function.

Livery The distinctive paint scheme (including lining, lettering and logos) applied to locomotives and coaches.

Low relief Models of buildings, generally placed along the backscene, which are of minimum depth.

Overhead collection The main system of full-sized electric railway current supply, where power is fed from an overhead wire supported by posts and gantries.

Pantograph The framework fitted on top of an electric locomotive or railcar to collect current from the overhead wire.

Pike (US) A model railroad layout.

Point *See* Turnout.

Power unit A black box unit containing transformer and rectifier to convert the mains voltage ac supply to low voltage dc for layout operation. It frequently incorporates an integral controller.

Prototype In modelling, the full-sized original on which the model is based.

Rail A shaped metal bar on which the wheels run. Most rails have a flat bottom, but in the UK, and parts of France, bullhead rail which had to be held in cast-iron chairs was used. This last type is obsolescent on main-line railroads. Full-sized rail is hot rolled from high-grade steel; model rail is drawn in nickel silver, brass and steel, the latter being coated to minimize rusting.

Railcar A self-propelled coach or set of coaches.

Rake The collective noun for a train of coaches.

Rapid transit A term, of US origin, used to describe urban railroad systems, generally electrified, providing an intensive passenger service, usually with no freight or parcel facilities. Since trains normally call at all stations, overall speeds are low by rail standards but the system is at least twice as fast as the competing road services.

Rollingstock In model railroad parlance, the collective term for coaches, wagons, vans and other non-powered vehicles.

Roundhouse A locomotive shed arranged around a turntable, with tracks radiating out from the table.

Scratchbuilder A modeller who relies on his own resources, rather than using kits.

DIRECTIONAL

A typical British goods yard during the steam age, full of loose-coupled short-wheelbase four-wheeled wagons which are catering for a much wider array of freight than is carried by rail today. In the foreground we have three cattle wagons alongside a set of stock pens on the cattle cock; to their right a closed van, used for general merchandise. The majority of wagons to the rear are 13-ton steel-sided mineral wagons, used in this instance for the carriage of coal, and once the staple earner of the British railroad network.

Scratchbuilding Constructing a model from raw materials and basic components, rather than from a complete kit.

Sleeper A beam placed under the rails to hold them to gauge. On the prototype these are usually of timber, but latterly concrete is gaining in favour. Pressed steel sleepers had considerable vogue but are now mainly used where termites are found. Plastic is used for the majority of commercial model sleepers.

Stud contact A system of current collection where power is collected from a series of studs down the centre of the track by means of a skate collector on the locomotive. This is used by Märklin HO as a replacement for the obsolete centre third.

Switch (1) An electrical device which interrupts or alters a circuit.
(2) In the US, the junction between two tracks. *See* Turnout.

Third rail (1) Prototype. A system of electrification where the current is collected from a third rail, generally located on the outside of the track. Mostly used on local rapid transit systems, but an extensive main-line network exists in the south of the UK.
(2) Model. An obsolescent system of current collection using a third rail to supply current. In commercial systems the rail was in the centre of the track, amateur usage placed it on the outside, as on the prototype.

Tie US term for sleeper.

Track Two rails, set to gauge on a series of sleepers (ties).

Tram(car) A rail-borne vehicle for urban transport, normally powered by electricity from an overhead wire and running along the streets. Extensively used in Europe, being revived in UK.

Tramway Originally another term for a railroad, it is now applied to urban rail systems which are laid on or alongside the streets. Now being revived in the UK under the term light railway.

Tram wire Tram systems have a single unsupported wire for pickup. Some narrow gauge systems have adopted this system for reasons of economy. Easy to install, it cannot be used where high speeds are required.

Turnout The fundamental unit of track formation, the place where two tracks join. Also called point (UK) and switch (US).

Two rail The normal method of model railroad electrification. Current is fed into one rail and returned through the other. Wheels must be insulated from each other to prevent short circuits.

Van (UK) A closed, four-wheeled goods vehicle.

INDEX

PICTURE CREDITS

The publishers would like to thank the following for supplying equipment and pictures for use in this book.

r = right; l = left; t = top; m = middle; b = bottom.

Continental Modeller: 2, 45, 58, 71t, 84. Hadley Hobbies: 3, 24, 26t, 26b, 28t, 28b(2), 29b. Brian Monaghan, *Continental Modeller*: 6, 10t, 13t, 17, 18, 20b, 38r, 54, 69, 83b, 91. C.J. Freezer: 7, 14, 16, 27, 41, 46, 48, 51, 52, 57, 62, 63, 73, 74, 75, 76, 77, 79, 87t, 90b. Brian Monaghan, *Railway Modeller*: 8, 10b, 13m, 13b, 26m, 28m, 29m, 38l, 39t, 39b, 43, 53b, 55, 60, 67, 68, 70, 71b, 72, 80, 81, 82, 86, 87b, 88, 90t, 94. *Model Railroader*: 9, 11, 12, 15, 19, 20t, 25, 29t, 31, 33, 37, 49, 53t, 56, 64–5, 78, 83t, 85, 89, 93.

Jacket: Models supplied by Hadley Hobbies; photography by Walter Gardiner.